3D Printing Technologies

3D Printing Technologies

Editor

Roberto Bernasconi

MDPI • Basel • Beijing • Wuhan • Barcelona • Belgrade • Manchester • Tokyo • Cluj • Tianjin

Editor
Roberto Bernasconi
Dipartimento di Chimica,
Materiali e Ingegneria Chimica
"Giulio Natta"
Politecnico di Milano
Italy

Editorial Office
MDPI
St. Alban-Anlage 66
4052 Basel, Switzerland

This is a reprint of articles from the Special Issue published online in the open access journal *Technologies* (ISSN 2227-7080) (available at: https://www.mdpi.com/journal/technologies/special_issues/3D_Printing_Technologies).

For citation purposes, cite each article independently as indicated on the article page online and as indicated below:

LastName, A.A.; LastName, B.B.; LastName, C.C. Article Title. *Journal Name* **Year**, *Volume Number*, Page Range.

ISBN 978-3-0365-3171-7 (Hbk)
ISBN 978-3-0365-3170-0 (PDF)

Contents

About the Editor

Roberto Bernasconi

Roberto Bernasconi earned his M. Sc. in Materials Engineering at Politecnico di Milano in 2011. From 2014 to 2018, he was a Ph.D student studying Materials Engineering. He earned his degree (PhD cum laude) for his thesis "Additive manufacturing and metallization for 3D microfabrication of functional devices". From 2018 to 2021, he worked as post-doctoral research fellow in Politecnico di Milano. Roberto Bernasconi is currently an Assistant Professor at Politecnico di Milano and runs the course "Electrochemical Energy Generators". His research fields include general and applied electrochemistry, and he has a specific interest in applications in the field of additive microfabrication. In particular, his main interest is bridging additive microfabrication (mainly in the forms of 3D printing and inkjet printing) with wet metallization technologies and surface modification processes in general. He is also active in the research of innovative materials for photovoltaics and energy applications and in the development of chemiresistive gas sensors. Roberto Bernasconi published 55 Scopus indexed scientific papers in renowned international journals, with an h-index of 13 and 466 citations (as of 02-02-2022). He also authored two book chapters. RB participated in the following research projects: Hardalt (EU 606110), LightMe (EU 814552) and PureNano (EU 821431). He serves as a Guest Editor for MDPI's *Technologies* and as Reviewer Editor for *Frontiers*.

Preface to "3D Printing Technologies"

The family of technologies collectively known as additive manufacturing (AM) technologies, and often called 3D-printing technologies, is rapidly revolutionizing industrial production. AM's potential to produce intricate and customized parts starting from a digital 3D model makes it one of the main pillars for the forthcoming Industry 4.0. AM also offers the opportunity to precisely engineer the properties of printed parts, with optimal material usage and an application-tailored design. Finally, the great variety of 3D-printing techniques available makes the use of many different materials possible: metals, polymers, ceramics, biomaterials and even living cells.

Thanks to these advantages over traditional manufacturing methodologies, AM finds potential applicability in virtually all production fields. As a natural consequence of this, research in this field is primarily focused on the development of novel materials and techniques for 3D printing. The importance of AM in the scientific community is attested by the thousands of papers published under the corresponding keywords in the last few years. This topic also caught the attention of major scientific publishers, resulting in the dissemination of specialized journals and dedicated books.

This Special Issue of *Technologies*, titled "3D Printing Technologies", aims at promoting the latest knowledge on materials, processes, and applications for AM. It is composed of six contributions, authored by influential scientists in the field of advanced 3D printing. The first paper, by Dr. Statnik et al., investigates the mechanical properties of metallic parts 3D printed using selective laser melting. In the second contribution, Dr. Koske et al. compare the influence of different infill designs on the mechanical properties of polymeric objects realized via fused deposition modeling. The third contribution of this Special Issue, proposed by me in collaboration with my coauthors, fits into the topic of AM for biomedical applications. In particular, it describes the 3D printing and wet metallization of untethered microdevices carrying hydrogel layers for controlled drug release. Additionally, the fourth contribution, proposed by Dr. Leonardi et al., regards the applicability of AM to the biomedical field. Indeed, Dr. Leonardi and her coauthors review the use of 3D-printed models for the preoperative planning and surgery of the pathology known as pectus excavatum. The last two contributions of this Special Issue deal with a fundamental aspect of polymeric-based AM technologies: the possibility to metallize the surface of the objects obtained. Dr. Kołczyk-Siedlecka et al. follow a wet metallization approach to realize decorated catalysts for methanol electro-oxidation, while Dr. Romani et al. use a dry approach (physical vapor deposition) for the metallization of thermoplastic polymers and composites.

The audience of this Special Issue includes professors, graduate students, researchers, engineers and specialists working in the field of AM. Finally, as the Guest Editor of this Special Issue, I would like to acknowledge the great efforts made by the authors to produce the high-quality research here presented. It takes a considerable amount of time to write, submit and review the manuscripts and recognizing this work is fundamental.

Roberto Bernasconi
Editor

 *technologies*

Article

In Situ SEM Study of the Micro-Mechanical Behaviour of 3D-Printed Aluminium Alloy

Eugene S. Statnik [1], Kirill V. Nyaza [2], Alexey I. Salimon [1,*], Dmitry Ryabov [2] and Alexander M. Korsunsky [3,1]

[1] HSM Lab, Center for Energy Science and Technology, Skoltech, Moscow 121205, Russia; eugene.statnik@skoltech.ru (E.S.S.); a.korsunsky@skoltech.ru (A.M.K.)

[2] Light Materials and Technologies Institute, UC RUSAL, Moscow 121096, Russia; mobiad@yandex.ru (K.V.N.); dmitriy.ryabov2@rusal.com (D.R.)

[3] MBLEM, Department of Engineering Science, University of Oxford, Oxford OX1 3PJ, UK

* Correspondence: a.salimon@skoltech.ru

Abstract: Currently, 3D-printed aluminium alloy fabrications made by selective laser melting (SLM) offer a promising route for the production of small series of custom-designed support brackets and heat exchangers with complex geometry and shape and miniature size. Alloy composition and printing parameters need to be optimised to mitigate fabrication defects (pores and microcracks) and enhance the parts' performance. The deformation response needs to be studied with adequate characterisation techniques at relevant dimensional scale, capturing the peculiarities of micro-mechanical behaviour relevant to the particular article and specimen dimensions. Purposefully designed Al-Si-Mg 3D-printable RS-333 alloy was investigated with a number of microscopy techniques, including in situ mechanical testing with a Deben Microtest 1-kN stage integrated and synchronised with Tescan Vega3 SEM to acquire high-resolution image datasets for digital image correlation (DIC) analysis. Dog bone specimens were 3D-printed in different orientations of gauge zone cross-section with respect to the fast laser beam scanning and growth directions. This corresponded to the varying local conditions of metal solidification and cooling. Specimens showed variation in mechanical properties, namely Young's modulus (65–78 GPa), yield stress (80–150 MPa), ultimate tensile strength (115–225 MPa) and elongation at break (0.75–1.4%). Furthermore, the failure localisation and character were altered with the change in gauge cross-section orientation. DIC analysis allowed correct strain evaluation that overcame the load frame compliance effect and helped to identify the unevenness of deformation distribution (plasticity waves), which ultimately resulted in exceptionally high strain localisation near the ultimate failure crack position.

Keywords: RS-333 alloy; SLM 3DP; in situ SEM tensile testing; DIC analysis; *Ncorr*

Citation: Statnik, E.S.; Nyaza, K.V.; Salimon, A.I.; Ryabov, D.; Korsunsky, A.M. In Situ SEM Study of the Micro-Mechanical Behaviour of 3D-Printed Aluminium Alloy. *Technologies* **2021**, *9*, 21. https://doi.org/10.3390/technologies9010021

Received: 12 February 2021
Accepted: 12 March 2021
Published: 15 March 2021

Publisher's Note: MDPI stays neutral with regard to jurisdictional claims in published maps and institutional affiliations.

1. Introduction

Following a period of rapid development since the early 2000s, the additive CAD/CAM technology of selective laser melting (SLM) 3D printing of metal alloys was expanded to include the use of aluminium alloys in the 2010s [1–3]. Important process aspects such as the alloy composition, laser scanning rate and post-processing parameters were systematically investigated to achieve desirable mechanical performance [4–6]. Developed on the basis of the widely used near-eutectic casting Al-Si alloys, the newly formulated 3D-printable Al-Si-Mg alloys facilitate the fabrication of parts with low porosity through ensuring extended solidification times and good flowability, leading to superior casting and printability behaviour and low susceptibility to hot cracking. The improvement in the liquid-phase viscosity by means of doping Al-Zn-Mg-Cu high-strength alloys with secondary alloying elements (Zr, Mn, Fe, Co and others) reduces the propensy of undesired defect assemblies (pores and coarse dendrite grains) down to acceptable levels [4]. Moreover, transition elements such as Zr, Ti or Mn have strong grain-refining effects, leading to significant improvements in the resistance to hot cracking due to the decreased size of dendrites, which allows better liquid metal support within solid–liquid regions during solidification. The introduction of such micro-alloying elements even in the form of nano-sized particles is considered to be an efficient mean of improving printability [4]. Nevertheless, other phenomena must be taken into consideration that occur at hierarchically different dimensional structural levels. These include metallurgical aspects (formation of oxide inclusions, liquation and inherited component inhomogeneity) and thermo-mechanical processes (cracking and residual stresses caused by strong thermal gradients and solidification shrinkage). Control over these phenomena is required in order to optimise the mechanical and functional performance of 3D-printable aluminium alloys and components to attain high geometric fidelity of 3D-printed parts [7].

Technologically, the more complex printable aluminium alloys present a number of advantages with respect to their traditional counterparts for niche applications where longer production time per article is tolerable. Prototypes, single units or small batches of miniature parts having complex geometry for use in fine mechanics applications, computer components and gadget hardware and robotics currently represent a clear scope for SLM technology evolution [8].

Additive manufacturing technologies can find application when specific combinations of properties are sought, such as for miniature heat exchangers of least mass or volume when both high strength and thermal conductivity are required. The performance indices to be minimised, according to Ashby [9], are $\frac{\rho}{\sigma_y^2 \cdot \lambda}$ and $\frac{1}{\sigma_y^2 \cdot \lambda}$, respectively, where ρ is density, σ_y is yield strength and λ is the thermal conductivity.

Traditionally, components such as heat exchangers are fabricated of 1XXX, 3XXX (Al-Mn) or 6XXX (Al-Mg-Si) series alloys. On the other hand, new aluminium alloys purposefully optimised for 3D printing possess both higher strength (compared to traditional AlSi$_{10}$Mg alloy used in 3D printing) and improved thermal properties. Though aluminium-based materials are generally expected to have high thermal conductivity [10], favourable to produce various parts for heat exchangers, 3D-printable AlSi$_{10}$Mg alloy manifests moderate heat conductivity. Moreover, 6XXX alloys possess rather high strength and acceptable thermal conductivity, as shown in Figure 1, but alloys such as 6061 are prone to cracking during the printing process [11], prompting researchers to seek new alloy formulations to adopt for AM processes. The composition of the 3D-printable RS-333 alloy from the Al-Mg-Si family considered in this paper was tuned by design to improve castability and thereby to provide a substitute for 6061 aluminium alloy in heat exchange applications.

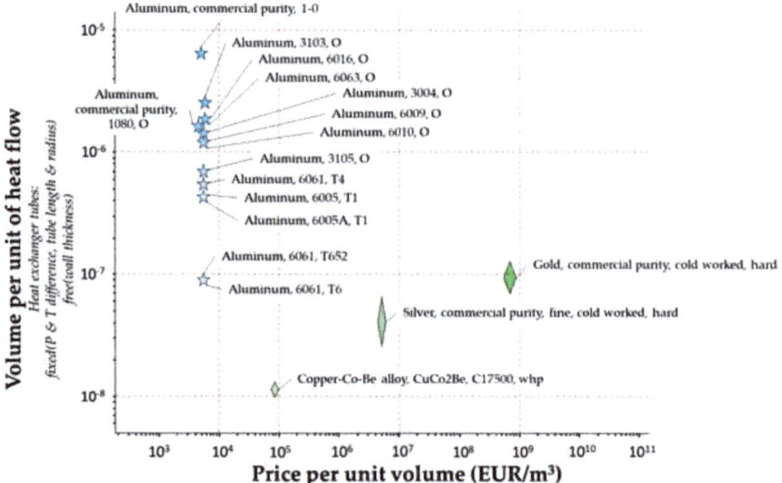

Figure 1. Comparison of some materials used in heat exchanging applications. Chart and data from CES EduPack 2019, Granta Design Limited, Cambridge, UK, 2019 [12].

Mechanical microscopy methods that couple optical and/or electron microscopy with mechanical testing and advanced digital image analysis have been rapidly developing in recent decades [13–17]. These methods are particularly suitable for in situ mechanical testing of near-net-shape miniature and thin parts inside an SEM chamber, since the deformation and fracture behaviour can be visualised and studied at high resolution, readily applicable to digital image correlation (DIC) analysis to reveal the peculiarities of deformation at the micrometre scale. Moreover, SLM 3D-printing of miniature fine mechanics parts having slim cross-sections occurs under the cooling and solidification conditions that differ from those for conventional engineering parts.

We report our findings in a systematic study of the micro-mechanical behaviour of tensile samples made from 3D-printable alloy RS-333. The acquisition of high-resolution SEM images was synchronised with in situ tensile testing of dog bone samples 3D-printed in different relative orientations of the laser scanning and growth directions with respect to the sample shape. DIC analysis was used to map the strain distribution and thus to trace the localisation of strain in the vicinity of the major crack. The interpretation of the experimental findings suggests a correlation between the duration of the effective cooling period for elementary added material volumes, on the one hand, and the mechanical performance on the other.

2. Materials and Methods

The powder of RS-333 (Al-3Si-0.5Mg) alloy was supplied by "Valcom-PM" Ltd. (Volgograd, Russia). The powder was produced by a nitrogen atomisation method. Particle size varied in the range of 20–63 μm with D50 = 42 μm according to the data from laser particle size analysis performed with ANALYSETTE 22 Nano Tec (Fritsch, Idar-Oberstein, Germany). The SEM appearance of the RS-333 powder is presented in Figure 2.

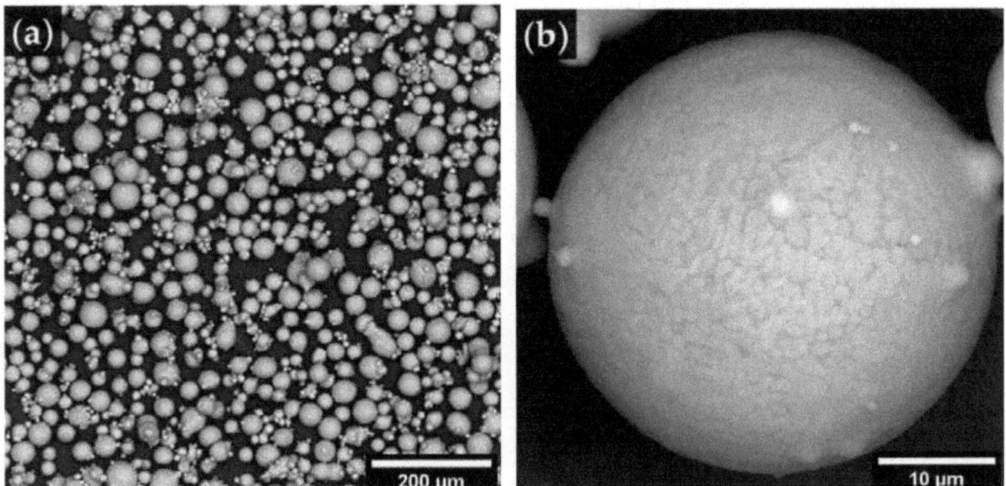

Figure 2. SEM images of RS-333 powders: (**a**) general view and (**b**) single particle microstructure.

A well-known challenge in SLM concerns 3D-printing of highly reflective aluminium alloy powders that possess inherent reflectivity of up to 90% in the IR range at ~1-μm wavelength. This was successfully overcome by the feedstock supplier using proprietary methods of powder surface modification (roughening) to allow this additive manufacturing technology to be applied at industrial scale.

The 3D-printing SLM process (powder bed fusion) was carried out using an EOS M290 SLM printer (Germany) equipped with a 400-W Yb fibre laser unit with a wavelength of 1075 nm. Argon of high purity was used during printing to avoid oxidation of the powder and the melt. Scanning speed used was 800 mm/s at the laser power of 370 W. Layer thickness was set to 30 μm. A post-printing heat treatment (aging) at 160 °C for 12 h to release residual stress and to promote precipitation hardening was applied to all printed samples. All samples were printed using "core" parameters without special skin exposure for further surface treatment.

In this research, ASTM E8 standard was used for sample geometry. Flat dog bone specimens having thickness of ca. 1 mm and gauge length of 10 mm were printed in a layer-by-layer process. Sets of at least 3 samples of each different orientation of main gauge axis with respect to the fast (X) and slow (Y) laser scanning directions and the growth axis (Z), as shown in Figure 3, were fabricated to study the influence of printing orientation on the mechanical response under tension. In our notation, the first character corresponds to the axis aligned with gauge length, the second to the axis aligned with gauge width. Supports used during printing were mechanically machined off to obtain samples of nominal dimensions.

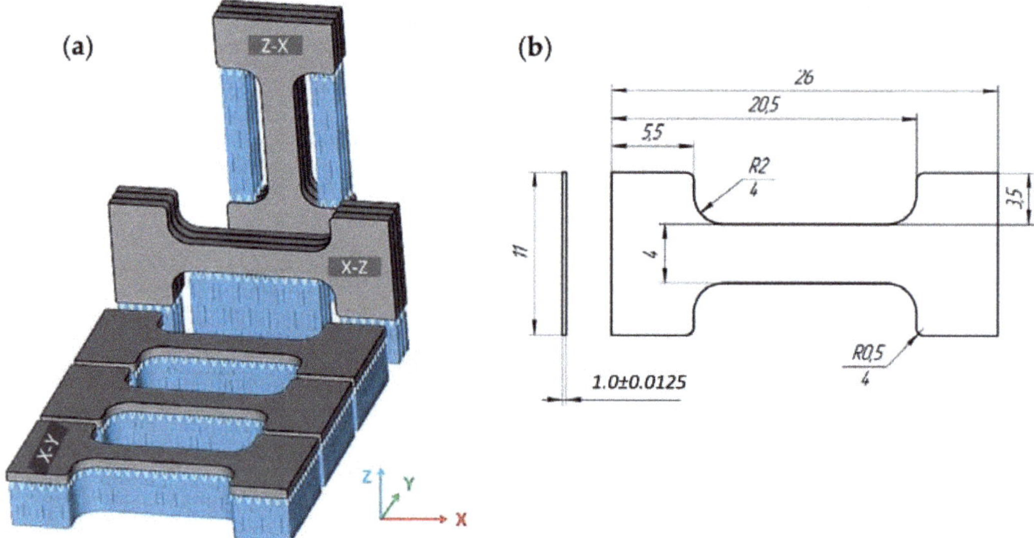

Figure 3. Test samples (**a**) of varying orientation during printing process, X-Y, X-Z and Z-X, shown from front to back, and (**b**) the nominal sample shape and dimensions.

Microstructure studies were performed after grinding and polishing using Struers laboratory equipment of as-printed sample cross-sections with no additional chemical etching. Optical microscope Zeiss Axio Observer 7 and scanning electron microscope Tescan MIRA 3 LMH (Tescan Company, Brno, Czech Republic) were applied to quantify residual pores and visualise internal fine structure, respectively. As a rule, the analysis of 10 fields of cross-sections in Z direction was carried out for porosity assessment.

In situ mechanical testing was facilitated through the use of a Deben Microtest 1-kN testing stage placed in the chamber of a Tescan Vega 3 SEM (Tescan Company, Brno, Czech Republic). The testing stage was operated under control of a Python code [18] to synchronise the mechanical loading (conducted at a permanent crosshead speed of 0.2 mm/min) with the acquisition of SEM images, as illustrated in Figure 4.

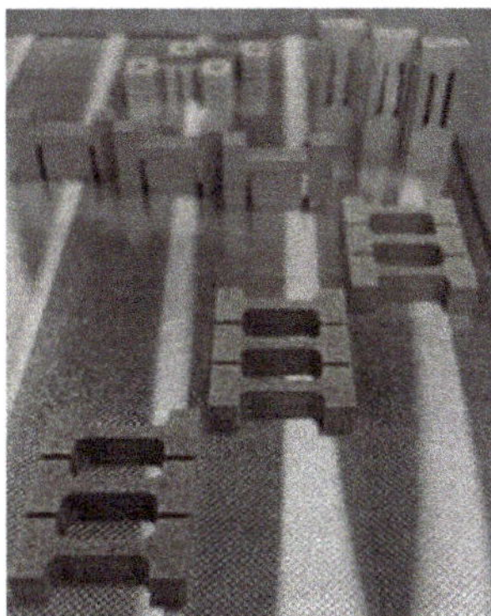

Figure 4. Appearance of 3D-printed specimens (**left**) and principal layout of experimental setup for in situ SEM study of mechanical response (**right**).

SEM images were acquired at the rate of 22 s per image in the secondary electron regime using 30 kV voltage and beam spot size of 400 nm. Videos S1–S6 of the specimen deformation and fracture process can be found in the Supplementary Materials.

Digital image correlation analysis with the use of open-source Matlab-based software *Ncorr* [19] was applied to the series of SEM images (up to ~40–50 images depending on the dataset acquired till sample break) to map the distribution of displacement and strain with subpixel resolution. The DIC algorithm compares two digital images (arrays containing digitalised intensity values) aiming to find the best match between pixel subsets. After the determination of the centre positions of corresponding pixel subsets, the displacement and eventually strain fields are calculated.

The definition of regions of interest and pattern quality (density of distinguishable surface features) affects both robustness and computing time in DIC analysis being performed with help of *Ncorr*. As seen in Figure 5, the pattern quality was satisfactory in terms of contrast and surface feature density. In this paper, 80% of the gauge area was used for DIC analysis, as illustrated in Figure 6.

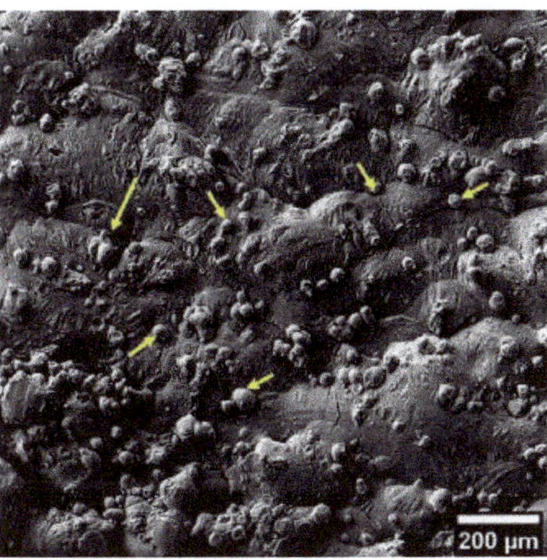

Figure 5. Appearance of surface of 3D-printed RS-333 alloy with the indicated (yellow arrows) surface features.

Figure 6. Illustration of typical ROI (Region Of Interest) selection for captured SEM image.

An important issue of elongation calibration at loading curves needs to be addressed when elastic moduli and true strains are sought to be derived from raw data acquired from a testing stage such as the Deben Microtest 1 kN. Crosshead displacement data are returned from a Linear Variable Differential Transformer (LVDT) sensor. Despite the high accuracy of the LVDT sensor, several factors must be taken into account when loading curves are analysed, namely (1) the effect of machine load frame compliance on the apparent deformation, and (2) the non-uniformity of the plastic and total strain (rate) during tensile deformation [20]. In the absence of suitable correction, the apparent deformation response of each sample depends on the stiffness of the test machine used. There are several methods of so-called specimen–machine coupling effect determination that allow the extraction of true specimen deformation data from the test results. The most straightforward method is the total deformation analysis method, in which the crosshead compliance correction for the displacement values is performed using the following formulae:

$$u_{total} = u_{sample} + u_{machine};\tag{1}$$

$$u_{sample} = u_{total} - u_{machine} = u_{total} - \frac{F}{k_{machine}},\tag{2}$$

where

u_{sample} is true sample displacement (mm);
u_{total} is the displacement measured during the test (mm);
F is the force measured during the test (N);
$k_{machine}$ is load frame stiffness (N/mm);
$1/k_{machine}$ is load frame compliance (mm/N).

In the present study, a special "stiff" Deben calibration specimen of SS316 steel with cross-section of 300 mm^2 and hence negligible deformation under load was used to obtain the calibration loading curve and derive $k_{machine}$ for the particular Deben Microtest 1-kN device.

On the other hand, DIC analysis is able to return strain maps in the region of interest (ROI) as well as the average strain along the main axis at the gauge length. The example of the selected ROI for each sample was similar in accordance with Figure 6. The longitudinal strain averaged over the ROI was used to plot the corresponding loading curve depicted in Figure 7.

In Figure 7, the data corrected for machine compliance in comparison with DIC analysis results are illustrated for a typical sample. It is apparent that DIC analysis yields the most reliable values of Young's modulus, so that, hereinafter, true strains determined by means of DIC analysis are used in the analysis.

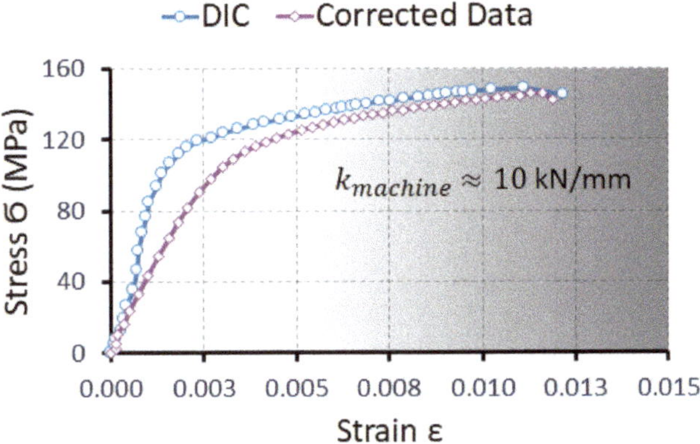

Figure 7. Illustration of sample stress–strain curves obtained from DIC analysis with the testing stage data corrected to account for the testing rig frame compliance.

3. Results and Discussion

3.1. Microstructure Studies

Selected parameters of the powder bed fusion process applied for the designed RS-333 powder result in the formation of a material structure with rather low porosity and having no detectable internal hot cracks, which indicates good service characteristics. Porosity calculated from the analysis of optical microscopy images as shown in Figure 8 is as low as 0.3%, which is typical for 3D-printed aluminium alloys [21].

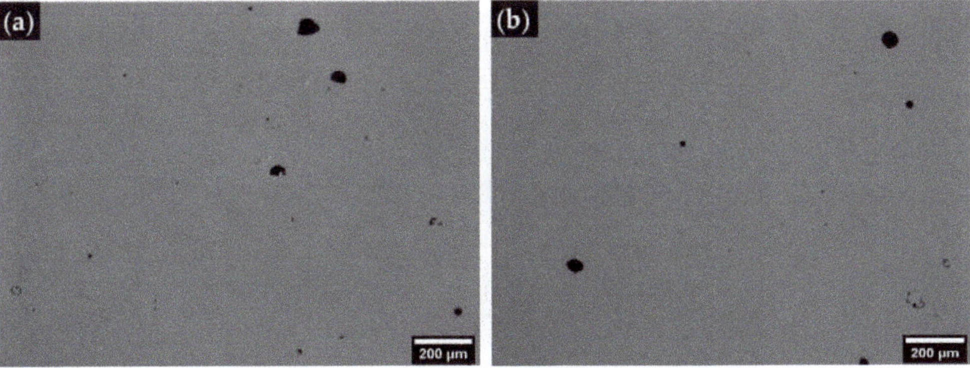

Figure 8. Representative images (**a,b**) of porosity within SLM RS-333 aluminium alloy parts taken in the X-Z sectional plane orientation.

SEM images of the structure formed in RS-333 alloy samples as a result of SLM 3D-printing and aging are represented in Figure 9. Si phase precipitates in the aluminium matrix are mainly represented by the curved and dashed chains, and some smaller equiaxial particles are also noticeable, suggesting that Si in RS-333 alloy is prone to some spheroidis-ation at the aging temperature applied. The improvement in heat conductivity takes place due to the decomposition of oversaturated solid solution. Aging temperature of 160 °C is, however, insufficient to complete the spheroidisation, eventually giving an optimal combination of high heat conductivity and mechanical performance.

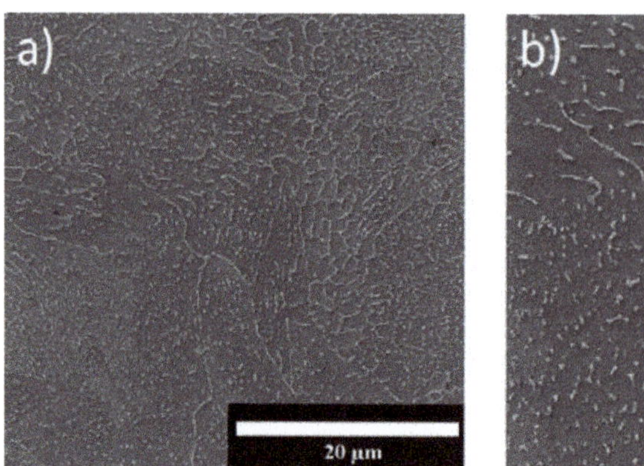
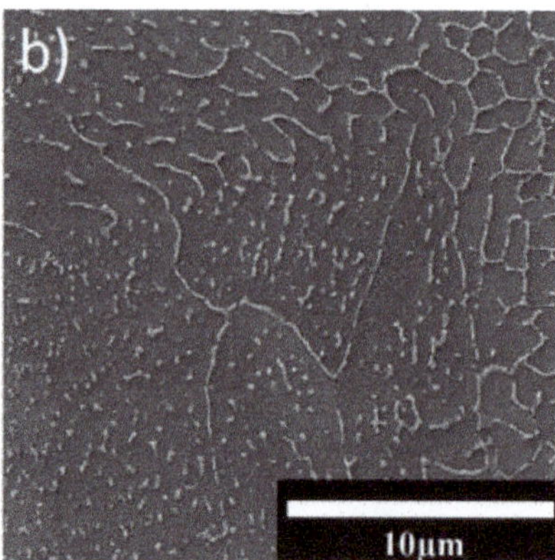

Figure 9. SEM images of RS-333 alloy microstructure following SLM printing and aging: (**a**) low and (**b**) high magnifications.

Surface appearance of an X-Z sample is represented in Figure 10, revealing relatively large (200–400 µm) and smooth clusters of molten material decorated with wrinkles and rare cracks, as well as the inclusion of non-molten powder particles. The rest of the supports are visible at the bottom edge of the gauge, where the clusters are apparently coarser and appear to be more separated with voids than in the middle zone of the gauge. These locations and support-free surfaces did not undergo stable continuous laser scanning and powder fusion. In contrast, the metallographic examination of the specimens' core (stable laser scanning and powder fusion) revealed residual porosity only (Figure 8) and no traces of unfused powder.

Figure 10. Surface appearance of X-Z sample with structure specification.

3.2. Mechanical Performance versus Sample Orientation

Mechanical performance is significantly affected by the growth orientation during SLM 3D-printing, as illustrated by the data represented in Figure 11 and Table 1. This technology is intrinsically complex and affects the structure (and mechanical performance) at hierarchically scaling—from sub-micrometre up to millimetre—dimensional levels, making the interpretation and understanding especially challenging in terms of intensive structure characterisation.

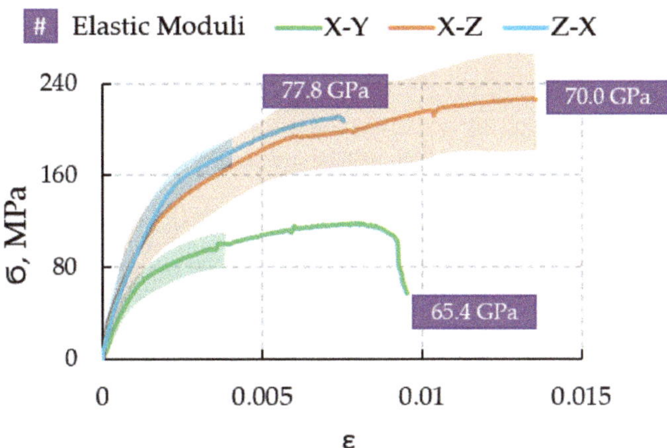

Figure 11. The typical stress–strain curves obtained for different sample printing orientations with the Young's modulus also indicated.

Table 1. Mechanical performance of RS-333 alloy samples with different sample orientations.

Sample Orientation	Young's Modulus, GPa	Yield Strength at 0.2% Strain, MPa	Ultimate Tensile Strength, MPa	Elongation at Rupture, %	Path Time, s	Layer Time, s
X-Y	65.4	81	115	0.95	0.03	0.30
X-Z [1]	70.0	132	227	1.36	0.03	0.10
Z-X	77.8	150	210	0.75	0.005	0.015

[1]—sample break takes place outside of gauge zone, i.e., in the rounded zone between gauge and clamp.

A conceptual model can be put forward to highlight the aspect of thermal history. Cooling rate from melting temperature and peak temperatures at reheating from subsequent paths and layers in a certain material micro-volume play an important role in the structure formation, including at least the following: (a) solid solution composition (and Si oversaturation) after solidification; (b) ultimate grain size; (c) development of aging processes—volume fraction and morphology of Si precipitates.

Taking into account that the area of the gauge cross-section directly depends on the sample orientation, one can easily estimate the characteristic times needed to print a path in plane (path time) and a layer (layer time) using the following formulae:

$$\tau_{path} = \frac{l_{path}}{v_{scan}}, \tag{3}$$

$$\tau_{lay} = \frac{S_{cs}}{d_{spot} \cdot v_{scan}}, \tag{4}$$

where τ_{path}—path time (s)

τ_{lay}—layer time (s);

l_{path}—path length (along fast scanning direction) in gauge zone (mm);

S_{sc}—area of cross-section in gauge zone (mm^2);

d_{spot}—diameter of fusion zone (scaled with laser power) (mm);

v_{scan}—scanning speed (mm/s).

The estimations listed in Table 1 were calculated for $d_{spot} = 300$ μm and $v_{scan} = 800$ mm/s in this research. The cooling rate for a particular micro-volume correlates with its particular position inside the cross-section as well as the geometry and area of the solidified cross-section, since both temperature gradient and the area of surrounding heat sinking zone depend on the latter characteristics. More accurate calculations of cooling rates can be modelled with a multi-physics FE simulation [22], but for qualitative considerations, cooling rate can be taken as a reverse function of path and layer time. Thus, the X-Y orientation with the highest cooling rate and less frequent reheating events ultimately brings the studied alloy to the structure state corresponding to solid solution heat treatment followed by natural aging. The smallest grain size is also expected for this orientation.

In contrast, the Z-X sample orientation with the lowest cooling rate and frequent reheating events tends to form the structure of an annealed (with obvious recrystallisation) or overaged solid solution with coarse grains. The X-Z orientation was found to be intermediate (and, perhaps, optimal) in terms of structure (solid solution heat treatment and artificial aging) and mechanical performance.

On the other hand, phenomena of another nature, such as the generation of residual stresses, chemical inhomogeneity and dimensional unevenness, are to be taken into account as factors which may predominate over microstructural effects.

This conceptual modelling may satisfactorily explain the correlation between SLM 3D-printing build orientation and mechanical performance attributes such as hardness, yield and ultimate tensile strength and elongation at rupture. These properties are mainly dominated by phase composition and structure morphology—grain size and orientation (texture). Few recent studies have been focused on the systematic analysis of aluminium [21], nickel [23], titanium alloys [24] and stainless steel [25].

Variations in mechanical properties against build-printing orientations were observed in many articles for different aluminium-based alloys prepared by the SLM technique [26]. For instance, Li X. [27], Ch S.R. [28] and Tang M. [29] investigated an $AlSi_{10}Mg$ alloy and achieved the same results as in our study; namely, samples manufactured in the vertical direction (Z-X) showed higher values of ultimate tensile strength than specimens built in the horizontal direction (X-Y). However, there are other studies where samples printed with equal conditions and composition demonstrated the inverse relation [30,31]. Moreover, this stochastic behaviour of mechanical characteristics was found during tensile tests of samples with other alloy compositions such as $AlSi_{10}$, A356 ($AlSi_7Mg_{0.3}$) and A357 ($AlSi_7Mg_{0.7}$) [27].

The issue of Young's modulus is more challenging to understand for aluminium alloys since these materials are almost elastically isotropic, so that the directional dependence of stiffness cannot be explained by texture variation. However, it should be noted that, similarly to our results, it has been shown that Young's modulus varies significantly within the range of 63–72 GPa for SLM 3D-printed $AlSi_{10}Mg$ alloy [31].

We believe that a strong variation in the elastic modulus for Al-Si-Mg is likely to be related to the peculiarities of the spatial arrangement of precipitates and nanometre-sized pores. However, detailed elucidation of these intricate relationships requires systematic studies by means of high-resolution techniques such as EBSD and FIB tomography, which are currently being performed by different research groups, including the present authors, in order to elucidate the contribution of grain orientation and porosity to the elastic behaviour of RS-333.

3.3. DIC Insight into Strain Distribution

It is demonstrated in Figures 12 and 13 and Figure A1 that DIC analysis mapping reveals the inhomogeneous distribution of strains even at the earliest stage of tension when a sample is macroscopically elastic. Strain distribution can be described as a wave with a wavelength along longitudinal axes of approximately 2 mm, which may suggest the influence of technological factors to be optimised for more strict quality control.

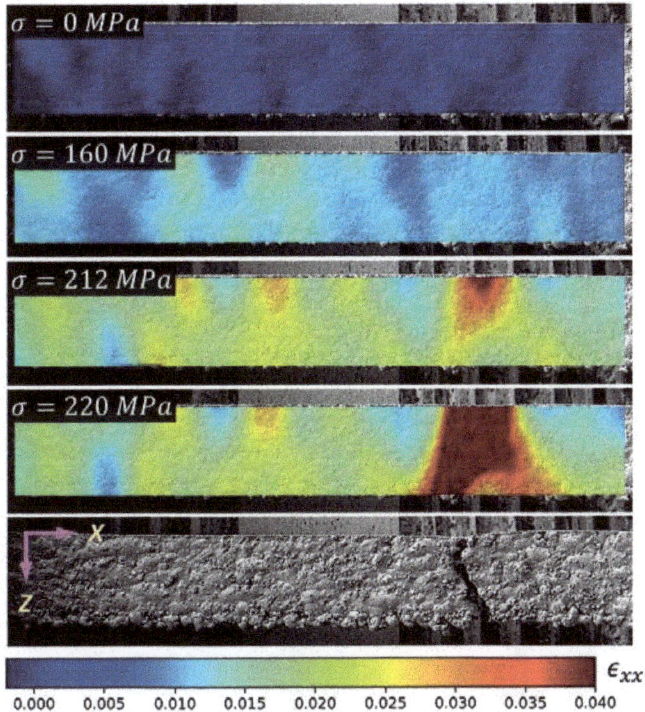

Figure 12. Demonstration of 2D DIC processing technique, namely continuous strain localisation tracking.

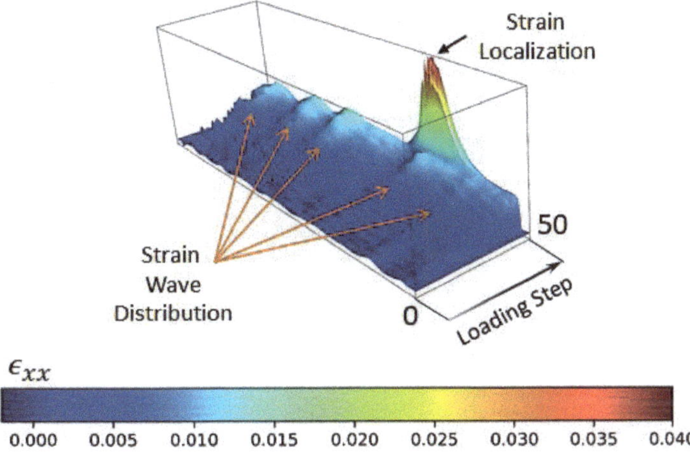

Figure 13. 3D DIC result representation with indicated strain wave-like distribution behaviour.

Localisation of strain in the vicinity of future cracks happens drastically and immediately before the break, which may be explained in stochastic terms, such as an inherited crack, a pore or overgrown grain or a non-metallic inclusion. In any case, almost no necking was noticed, leaving the issue of survivability unresolved.

It is worth noting that some samples having X-Z built orientation showed that the localisation of strains and further rupture occur in the rounded zone between the gauge and clamp. In this zone, the transition from long gauge zone to short clamp zone takes place at longitudinal laser scanning. There is a narrow zone where abrupt changes in path length, layer time and cooling rate occur, causing the unevenness of the temperature field and residual stresses at further overall cooling (temperature conditions are to be simulated with computer modelling to quantify this effect). We suggest that the temperature and time of post-processing annealing were not optimised for the release of these stresses. This gives very useful guidance for good practice in SLM 3D-printing.

3.4. Fracture Surface Appearance

The appearance of the fracture surface in Figure 14 suggests that the RS-333 alloy manifests some ductility during the initiation and growth of a crack. The crack line occurs at an angle around 60 degrees to the longitudinal axis; however, smaller and larger angles were also detected.

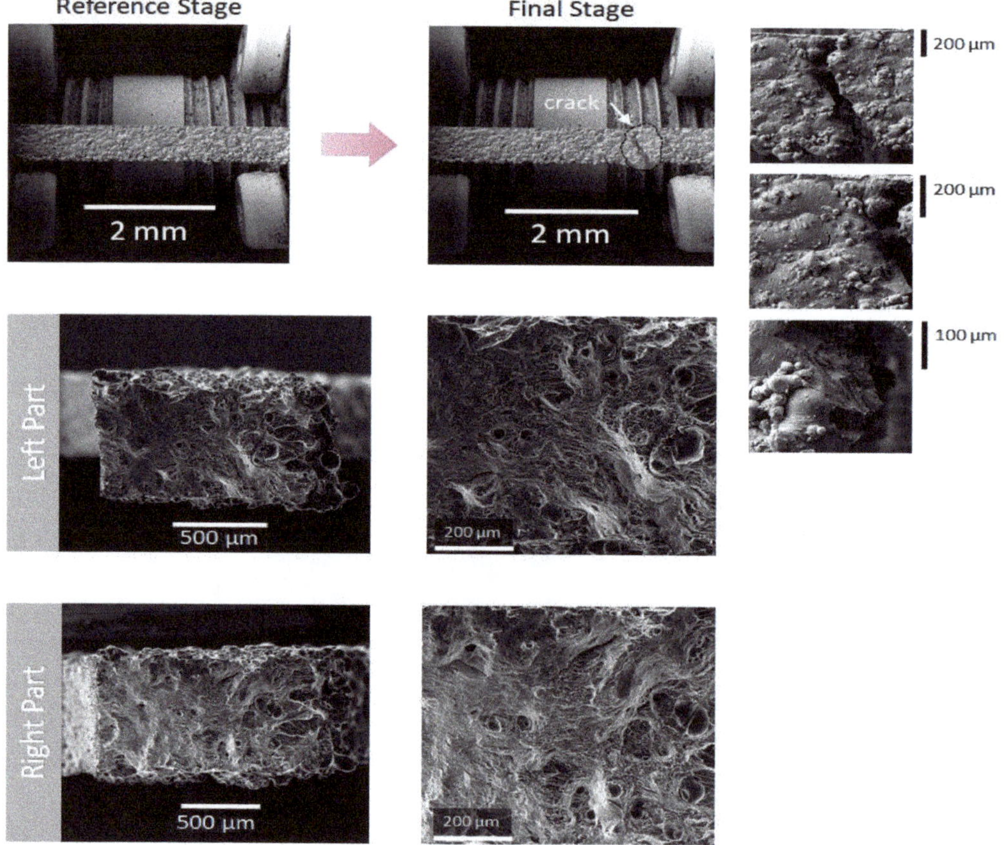

Figure 14. Appearance of fracture surface of an X-Z specimen.

The fracture surface (Figure 14, from left to right) has relatively smooth zone with few round pits, followed by a zone with elongated fringe features and finally a zone with coarse pit features at the edge, where the rest of the supports were noticed (Figure 10). It can be seen in Figure 12 that the crack is initiated from the flat edge zone opposite to the edge with coarse clusters and the rest of the supports. It seems that the crack starts to grow in accordance with ductile mechanisms (smooth zone) and terminates as a brittle crack (rough zone with elongated fringe features) when substantial stress concentration is reached. The initiation of a crack happens in the zone where sample growth ends and material consolidation evolves, with no remelting and reheating. On the other hand, this zone has a much firmer substrate (specimen body) than at the opposite stage, where only supports are present. Deeper structural investigations are required to reliably identify the main physical phenomena governing the fracture behaviour.

4. Conclusions

For dog bone specimens, SLM 3D-printed Al-Si-Mg alloy RS-333 reveals a significant interrelation between the specimen orientation and mechanical tensile performance, hinting at a practical approach for the optimisation of the overall performance of heat exchanger articles. An X-Z specimen orientation shows the highest values of yield and ultimate strength and elongation at break at a Young's modulus of approximately 70 GPa. in situ SEM studies of tension response and corresponding DIC analysis are particularly suitable to highlight the peculiarities of mechanical behaviour, such as unevenness of strains and their localisation in the vicinity of the ultimate crack. The complexity of the processes (solidification at different cooling rates, frequency of reheating events, natural and artificial aging) forming the final structure requires fundamental characterisation research to guide the optimisation of overall performance in the most efficient way.

Supplementary Materials: The following are available online at https://www.mdpi.com/2227-708 0/9/1/21/s1, Videos S1–S6: Fracture process videos of specimens with different printing orientations.

Author Contributions: Conceptualisation, D.R. and K.V.N.; methodology, K.V.N. and E.S.S.; software, E.S.S.; validation, D.R., A.I.S. and A.M.K.; formal analysis, E.S.S. and A.I.S.; investigation, E.S.S.; resources, A.I.S.; data curation, E.S.S.; writing—original draft preparation, D.R., E.S.S., A.I.S., and A.M.K.; writing—review and editing, E.S.S., A.I.S., and A.M.K.; visualisation, E.S.S.; supervision, A.I.S. and A.M.K.; project administration, D.R.; funding acquisition, D.R. All authors have read and agreed to the published version of the manuscript.

Funding: This research received no external funding.

Institutional Review Board Statement: Not applicable.

Informed Consent Statement: Informed consent was obtained from all subjects involved in the study.

Data Availability Statement: The data presented in this study are available in the Supplementary Materials.

Acknowledgments: The authors thank Julia Malakhova for the sample preparation.

Conflicts of Interest: The authors declare no conflict of interest.

Appendix A

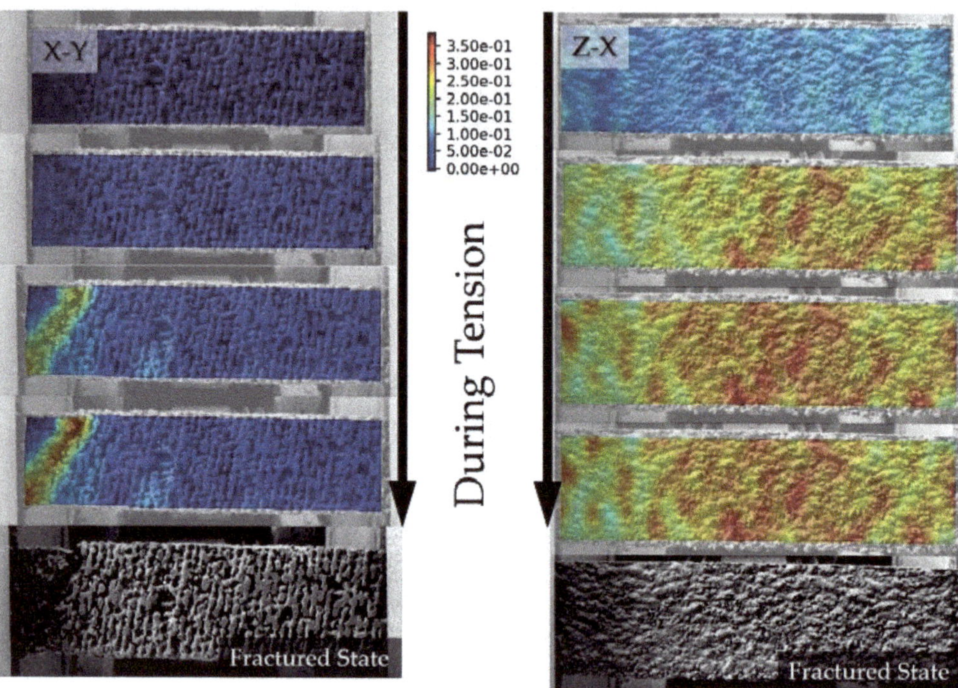

Figure A1. 2D strain distributions along loading axis for samples with different printing orientations: (**left**) X-Y and (**right**) Z-X.

References

1. Louvis, E.; Fox, P.; Sutcliffe, C.J. Selective laser melting of aluminium components. *J. Mater. Process. Technol.* **2011**, *211*, 275–284. [CrossRef]
2. Thijs, L.; Kempen, K.; Kruth, J.P.; Van Humbeeck, J. Fine-structured aluminium products with controllable texture by selective laser melting of pre-alloyed AlSi$_{10}$Mg powder. *Acta Mater.* **2013**, *61*, 1809–1819. [CrossRef]
3. Olakanmi, E.O.; Cochrane, R.F.; Dalgarno, K.W. A review on selective laser sintering/melting (SLS/SLM) of aluminium alloy powders: Processing, microstructure, and properties. *Prog. Mater. Sci.* **2015**, *74*, 401–477. [CrossRef]
4. Martin, J.H.; Yahata, B.D.; Hundley, J.M.; Mayer, J.A.; Schaedler, T.A.; Pollock, T.M. 3D printing of high-strength aluminium alloys. *Nature* **2017**, *549*, 365–369. [CrossRef]
5. Aboulkhair, N.T.; Everitt, N.M.; Ashcroft, I.; Tuck, C. Reducing porosity in AlSi$_{10}$Mg parts processed by selective laser melting. *Addit. Manuf.* **2014**, *1*, 77–86. [CrossRef]
6. Read, N.; Wang, W.; Essa, K.; Attallah, M.M. Selective laser melting of AlSi$_{10}$Mg alloy: Process optimisation and mechanical properties development. *Mater. Des.* **2015**, *65*, 417–424. [CrossRef]
7. Imonelli, M.; Aboulkhair, N.; Rasa, M.; East, M.; Tuck, C.; Wildman, R.; Salomons, O.; Hague, R. Towards digital metal additive manufacturing via high-temperature drop-on-demand jetting. *Addit. Manuf.* **2019**, 30. [CrossRef]
8. Salimon, A.; Brechet, Y.; Ashby, M.F.; Greer, A.L. Selection of applications for a material. *Adv. Eng. Mater.* **2004**, *6*, 249–265. [CrossRef]
9. Ashby, M.F. *Materials Selection in Mechanical Design*, 4th ed.; Butterworth-Heinemann: Oxford, UK, 2010; p. 640.
10. Klemens, P.G.; Williams, R.K. Thermal conductivity of metals and alloys. *Int. Met. Rev.* **1986**, *31*, 197–215. [CrossRef]
11. Senthamarai, C.K. *IOP Conferences Series: Materials Science and Engineering*; IOP Publishing: Bristol, UK, 2020; p. 988.
12. *CES EduPack Software*; Granta Design Limited: Cambridge, UK, 2019; Available online: https://www.ansys.com/products/materials (accessed on 10 February 2021).
13. Hanon, M.M.; Marczis, R.; Zsidai, L. Anisotropy Evaluation of Different Raster Directions, Spatial Orientations, and Fill Percentage of 3D Printed PETG Tensile Test Specimens. *Key Eng. Mater.* **2019**, *821*, 167–173. [CrossRef]

14. Sui, T.; Salvati, E.; Zhang, H.; Nyaza, K.; Senatov, F.S.; Salimon, A.I.; Korsunsky, A.M. Probing the complex thermo-mechanical properties of a 3D-printed polylactide-hydroxyapatite composite using in situ synchrotron X-ray scattering. *J. Adv. Res.* **2018**, *16*, 113–122. [CrossRef]
15. Statnik, E.S.; Salimon, A.I.; Korsunsky, A.M. On the application of digital optical microscopy in the study of materials structure and deformation. *Mater. Today: Proc.* **2020**, *33*, 1917–1923. [CrossRef]
16. Statnik, E.S.; Dragu, C.; Besnard, C.; Lunt, A.J.G.; Salimon, A.I.; Maksimkin, A.; Korsunsky, A.M. Multi-Scale Digital Image Correlation Analysis of In Situ Deformation of Open-Cell Porous Ultra-High Molecular Weight Polyethylene Foam. *Polymers* **2020**, *12*, 2607. [CrossRef]
17. Statnik, E.S.; Ignatyev, S.D.; Stepashkin, A.A.; Salimon, A.I.; Chukov, D.; Kaloshkin, S.D.; Korsunsky, A.M. The Analysis of Micro-Scale Deformation and Fracture of Carbonized Elastomer-Based Composites by In Situ SEM. *Molecules* **2021**, *26*, 587. [CrossRef] [PubMed]
18. Blaber, J.; Adair, B.; Antoniou, A. Ncorr: Open-source 2D Digital Image Correlation Matlab software. *Exp. Mech.* **2015**, *55*, 1105–1122. [CrossRef]
19. Holbrook, J.A.; Swearengen, J.C.; Rohde, R.W. Specimen-test machine coupling and its implications for plastic-deformation models. *ASTM Int.* **1982**, 80–101. [CrossRef]
20. Nesma, T.; Aboulkhair, M.S.; Parry, L.; Ashcroft, I.; Tuck, C.; Hague, R. 3D printing of Aluminium alloys: Additive Manufacturing of Aluminium alloys using selective laser melting. *Prog. Mater. Sci.* **2019**, 106. [CrossRef]
21. Fernandez-Zelaia, P.; Kirka, M.M.; Dryepondt, S.N.; Gussev, M.N. Crystallographic texture control in electron beam additive manufacturing via conductive manipulation. *Mater. Des.* **2020**, 195. [CrossRef]
22. De Jager, B.; Zhang, B.; Song, X.; Papadaki, C.; Zhang, H.; Brandt, L.R.; Salvati, E.; Sui, T.; Korsunsky, M.A. Texture and Microstructure Analysis of IN718 Nickel Superalloy Samples Additively Manufactured by Selective Laser Melting. In Proceedings of the International MultiConference of Engineers and Computer Scientists 2017 Vol II, IMECS 2017, Hong Kong, 15–17 March 2017.
23. Zhang, D.; Qiu, D.; Gibson, M.A.; Zheng, Y.; Fraser, H.L.; StJohn, D.H.; Easton, M.A. Additive manufacturing of ultrafine-grained high-strength titanium alloys. *Nature* **2019**, *576*, 91–95. [CrossRef]
24. Murr, L.E. A Metallographic Review of 3D Printing/Additive Manufacturing of Metal and Alloy Products and Components. *Met. Microstruct. Anal.* **2018**, *7*, 103–132. [CrossRef]
25. Ponnusamy, P.; Rashid, R.A.R.; Masood, S.H.; Ruan, D.; Palanisamy, S. Mechanical Properties of SLM-Printed Aluminium Alloys: A Review. *Materials* **2020**, *13*, 4301. [CrossRef] [PubMed]
26. Li, X.; Ni, J.; Zhu, Q.; Su, H.; Cui, J.; Zhang, Y.; Li, J. Structure and Mechanical Properties of the $AlSi_{10}Mg$ Alloy Samples Manufactured by Selective Laser Melting. *IOP Conf. Ser. Mater. Sci. Eng.* **2017**, *269*, 12081. [CrossRef]
27. Ch, S.R.; Raja, A.; Nadig, P.; Jayaganthin, R.; Vasa, N. Influence of working environment and built orientation on the tensile properties of selective laser melted $AlSi_{10}Mg$ alloy. *Mater. Sci. Eng. A* **2019**, *750*, 141–151. [CrossRef]
28. Tang, M.; Pistorius, P.C. Anisotropic mechanical behavior of $AlSi_{10}Mg$ parts produced by selective laser melting. *JOM* **2017**, *69*, 516–522. [CrossRef]
29. Rosenthal, I.; Stern, A.; Frage, N. Microstructure and mechanical properties of $AlSi_{10}Mg$ parts produced by the laser beam additive manufacturing (AM) technology. *Met. Microstruct. Anal.* **2014**, *3*, 448–453. [CrossRef]
30. Rosenthal, I.; Stern, A.; Frage, N. Strain rate sensitivity and fracture mechanism of $AlSi_{10}Mg$ parts produced by selective laser melting. *Mater. Sci. Eng. A* **2017**, *682*, 509–517. [CrossRef]
31. Awd, M.; Stern, F.; Kampmann, A.; Kotzem, D.; Tenkamp, J.; Walther, F. Microstructural Characterization of the Anisotropy and Cyclic Deformation Behavior of Selective Laser Melted $AlSi_{10}Mg$ Structures. *Metals* **2018**, *8*, 825. [CrossRef]

 *technologies*

Article

Infill Designs for 3D-Printed Shape-Memory Objects

Daniel Koske and Andrea Ehrmann *

Faculty of Engineering and Mathematics, Bielefeld University of Applied Sciences, 33619 Bielefeld, Germany;
daniel.koske@fh-bielefeld.de
* Correspondence: andrea.ehrmann@fh-bielefeld.de

Abstract: Shape-memory polymers (SMPs) can be deformed, cooled down, keeping their new shape for a long time, and recovered into their original shape after being heated above the glass or melting temperature again. Some SMPs, such as poly(lactic acid) (PLA), can be 3D printed, enabling a combination of 3D-printed shapes and 2D-printed, 3D-deformed ones. While deformation at high temperatures can be used, e.g., to fit orthoses to patients, SMPs used in protective equipment, bumpers, etc., are deformed at low temperatures, possibly causing irreversible breaks. Here, we compare different typical infill patterns, offered by common slicing software, with self-designed infill structures. Three-point bending tests were performed until maximum deflection as well as until the maximum force was reached, and then the samples were recovered in a warm water bath and tested again. The results show a severe influence of the infill pattern as well as the printing orientation on the amount of broken bonds and thus the mechanical properties after up to ten test/recovery cycles.

Keywords: poly(lactic acid) (PLA); shape-memory polymer (SMP); fused deposition modeling (FDM); 3D printing; infill pattern

1. Introduction

Shape-memory polymers (SMPs) can recover their initial shape after deformations, triggered by an external stimulus, which is, in many cases, heat [1]. These properties are highly interesting for research and development and led to the development of many new polymers and polymer blends showing a shape-memory effect [2]. Typical applications can be found in spacecraft, biomedicine or smart textiles [3–5], with SMPs having different shapes such as bulk materials, films or foams [6–8].

Another possibility to create objects from SMPs is offered by 3D printing, e.g., stereolithography of fused deposition modeling (FDM) [9]. One of the materials most often used in 3D printing, poly(lactic acid) (PLA), shows such shape-memory properties [10]. It has, however, the disadvantage that it can only be elongated by ~10% until it breaks [11]. This leads to the question of how to avoid breaks by sophisticated constructions of samples.

Langford et al. used origami-inspired structures for this purpose and found that especially herringbone tessellated tubes could be strongly compressed and recovered afterwards, making them usable for biomedical scaffolds [12]. A simpler origami structure was suggested by Mehrpouva et al., who printed a flat structure foldable into a pyramid at higher temperatures [13].

More common structures were investigated by different groups, e.g., honeycomb, 3D honeycomb or gyroid structures, which are typical infill patterns, offered by many slicer programs [14–17]. Here, however, the general problem occurred that recovery after cold deformation was never perfect, with recovery ratios clearly below 100% due to undesired broken bonds.

Here, we perform three-point bending tests on test samples with different infill patterns, partly chosen from those offered by the slicer software, partly self-developed. Our results show that not only the infill pattern but also the printing orientation has a significant impact on the recovery properties and lead to suggestions on how to improve recovery after cold deformation.

Citation: Koske, D.; Ehrmann, A. Infill Designs for 3D-Printed Shape-Memory Objects. *Technologies* **2021**, *9*, 29. https://doi.org/10.3390/technologies9020029

Academic Editor: Roberto Bernasconi

Received: 27 March 2021
Accepted: 13 April 2021
Published: 16 April 2021

Publisher's Note: MDPI stays neutral with regard to jurisdictional claims in published maps and institutional affiliations.

2. Materials and Methods

The samples used in this study were printed using a MEGA-S FDM 3D printer (ANY-CUBIC; Shenzhen Anycubic Technology Co., Ltd., Shenzhen, China). With a nozzle diameter of 0.4 mm, a layer thickness of 0.2 mm was selected for the first layer and 0.12 mm for the other layers. The printing temperature was set at 200 °C and the heating bed temperature was set at 60 °C constant.

The test specimens made of PLA (GIANTARM PLA filament 1.75 mm, silver; no additives included to increase crystallinity [18]) were designed and printed with dimensions of 120 mm × 15 mm × 6 mm according to polymer test specimens (ISO 20753:2018). In order to investigate the influence of the filling pattern alone, no wall contours were printed, so the entire specimens were deflected only on the filling pattern and the lower and upper support plates of 1 mm each. The main dimensions were kept for the three-point bending test, and only the internal structure was changed.

The following support patterns and infill densities were applied to the specimens (Table 1): pattern line with 100% infill and 80% printing speed (LN100); pattern gyroid with 15% infill and 60% printing speed (GY15); pattern octet with 15% infill and 60% print speed (OC15); and self-designed filling structures in the form of hollow cylinders with 100% infill and 60% printing speed (ZHR100), a leaf spring construction (LP100) and a combination of the two (EW100). Some of the samples (ZHR100, LP100 and EW100) were printed lying flat on the printing bed as well as on the long edge. Table 1 shows all samples used here.

Table 1. Samples under examination in this study.

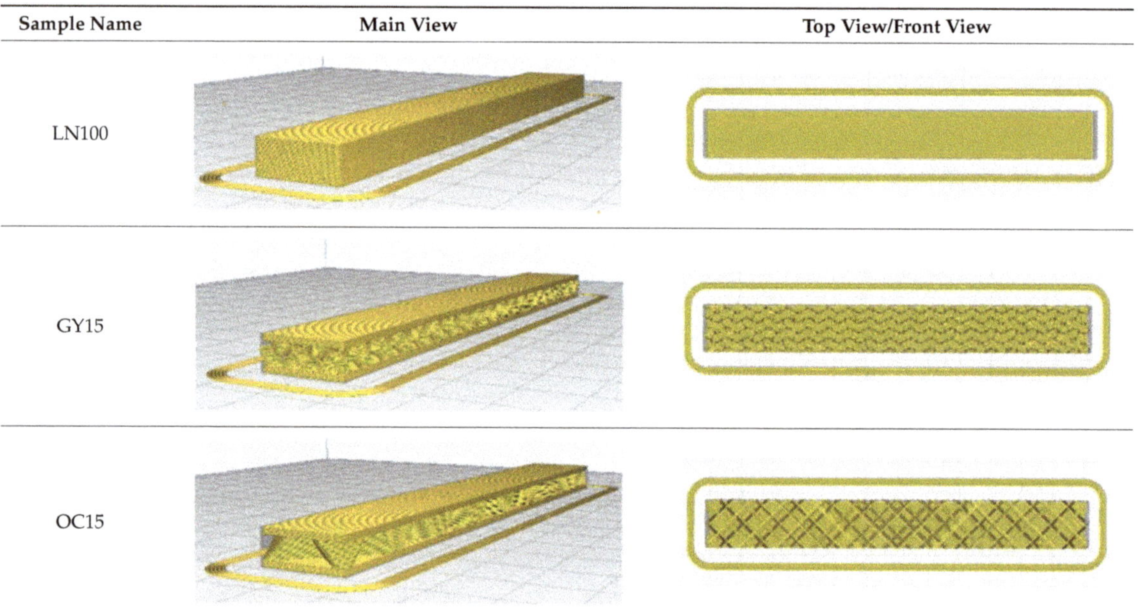

Sample Name	Main View	Top View/Front View
LN100		
GY15		
OC15		

Table 1. *Cont.*

Sample Name	Main View	Top View/Front View
ZHR100		
LP100		
EW100		

These samples were examined by a three-point bending test, using a universal testing machine (Kern & Sohn, Balingen-Frommern, Germany). Generally, a minimum of 4 specimens for each sample were examined, three of which were bent until breaking or until reaching the maximum deflection possible in the system. The fourth one was bent until the previously measured point of maximum force was reached and then relaxed and recovered in a water bath of (60 ± 1) °C for 1 min, before it was cooled down in another water bath at room temperature for 1 min. Since the aim of this investigation was avoiding broken areas, not optimization of the recovery process, these values were kept unchanged [19]. This temperature was chosen since amorphous PLA, as it typically is produced by 3D printing without a following heat treatment, has a glass transition temperature around 56 °C [20,21] which was verified in previous tests [22]. It should be mentioned that using filaments from different producers, and even using a filament of different color from the same producer, may lead to different results as most producers do not mention possible additives. In particular, so-called high-temperature PLA (HT-PLA) includes different additives to increase crystallinity and, correspondingly, the glass transition temperature [18], enabling even autoclaving at 121 °C without deformation of the sample [23].

For the optical examinations of the samples after the tests, a digital microscope, Camcolms2 (Velleman, Gavere, Belgium), was used.

3. Results and Discussion

Figure 1 depicts the three-point bending tests of the specimens shown in Table 1. For the self-designed infill patterns, samples were printed lying on the flat side as well as on the long edge.

As expected, the completely filled sample LN100 shows the highest maximum force. For the possible deflection of ~17.5 mm, none of them broke.

The gyroid infill was found advantageous in another study working with cubes in which a linear object was pressed till half of the original height [15–17]. Here, it shows, by far, the lowest maximum force. The sudden drop in the force around 5–7 mm deflection, visible in Figure 1b, is correlated with a delamination inside the infill, fully separating the top and bottom layers along one half of the respective specimen.

Sample OC15 shows a similar slope of the curve to LN100. Interestingly, the maximum force is approximately half the value found for LN100, while the infill was only chosen as 15%, showing that this infill performs better in relation to the sample mass than the completely filled sample.

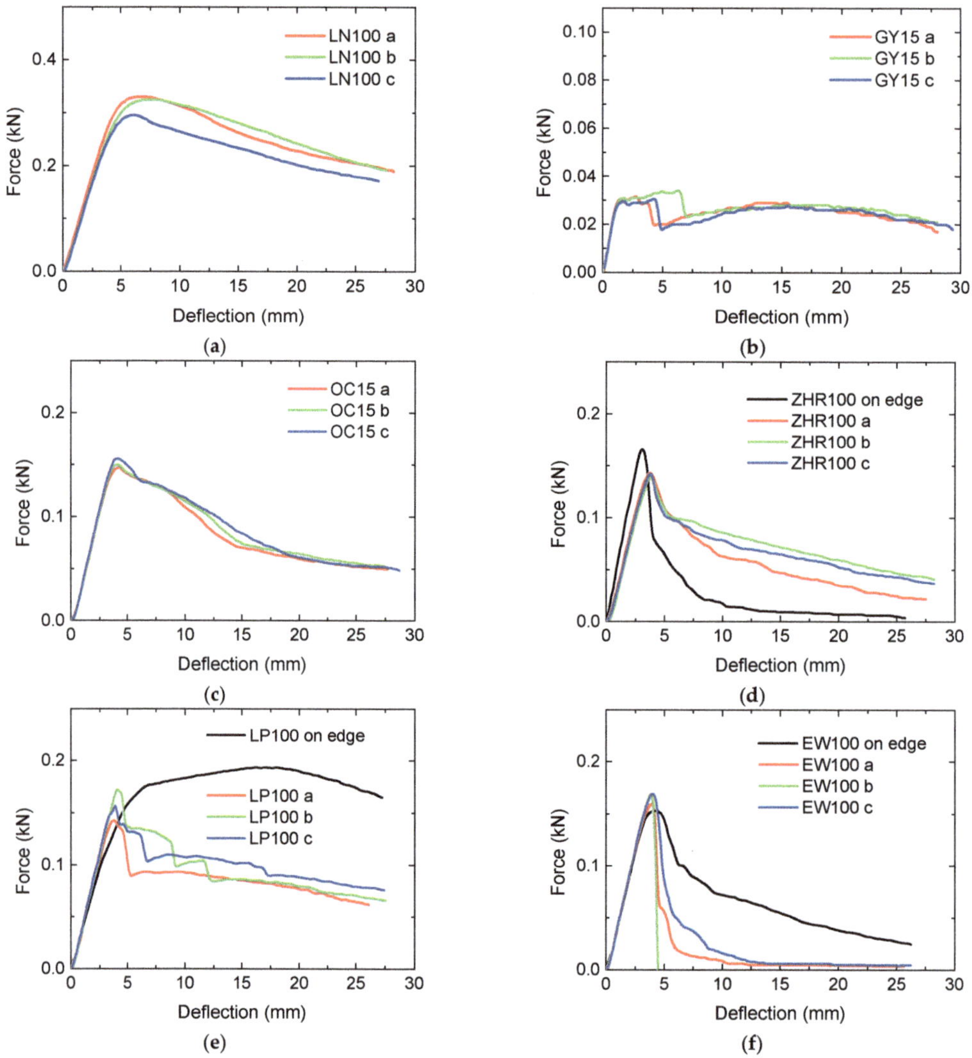

Figure 1. Three-point bending tests of samples with different infill patterns: (**a**) LN100; (**b**) GY15; (**c**) OC15; (**d**) ZHR100 including one specimen printed on the long edge; (**e**) LP100 including one specimen printed on the long edge; and (**f**) EW100 including one specimen printed on the long edge. Y-axis scales differ.

The samples with the first self-designed infill pattern, ZHR100, showed only a very narrow maximum, indicating that this infill will not work very well in repeated tests where deflection is planned to end at the point of maximum force. In all cases, the samples showed a nearly completely broken lower plate at the position of the bending line. Similar effects are visible for samples LP100 and EW100.

Comparing the samples printed on the long edge, this has a slightly higher, but even narrower, peak for ZHR100, making this sample even less suitable for recovery tests than the original printing direction. For sample EW100, the opposite effect occurs. For sample LP100, however, a completely different shape is visible. Here, a very broad maximum is found, making this sample printed on the edge highly interesting for recovery tests. Due to the width of this curve, tests can be performed for a broad range of deflections.

The corresponding microscopic images of the samples after the bending tests (always specimen "a" in Figure 1) are shown in Figure 2.

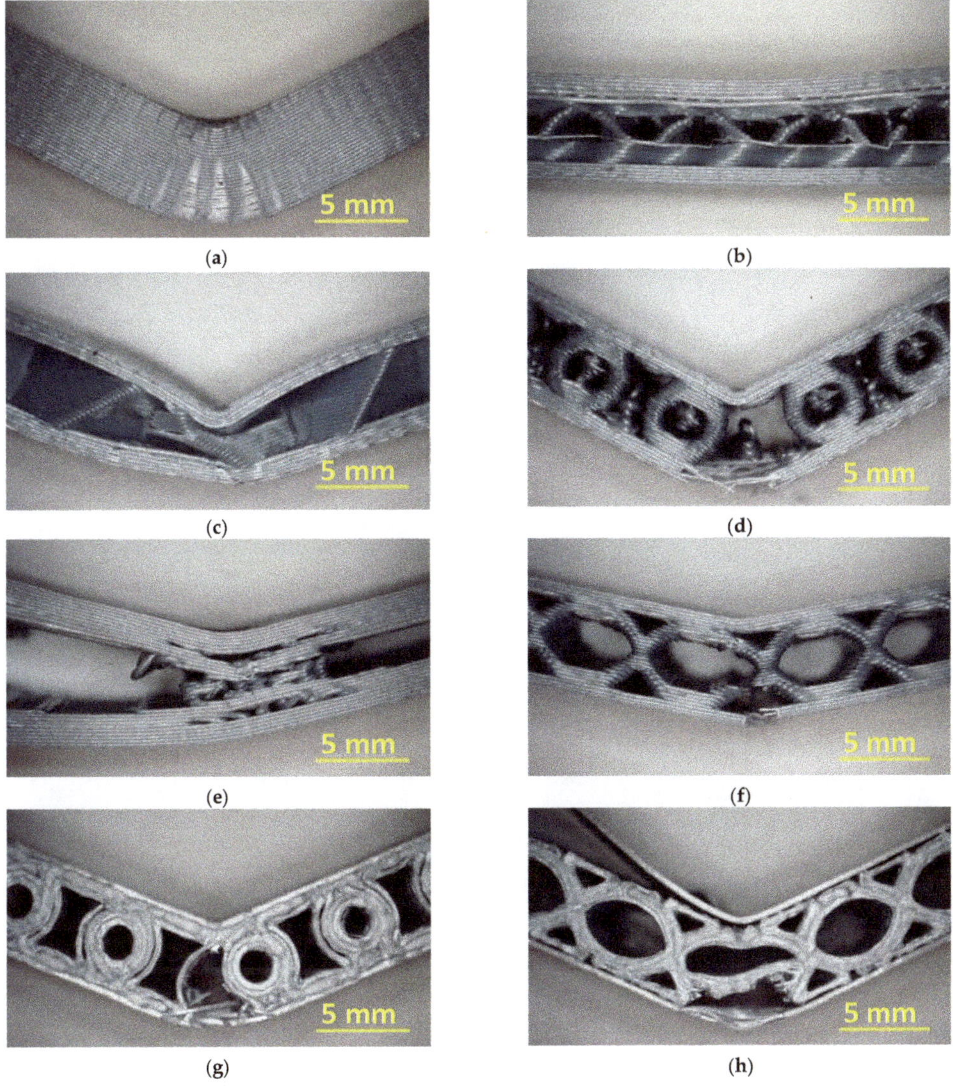

Figure 2. Microscopic images taken after three-point bending tests of samples with different infill patterns: (**a**) LN100; (**b**) GY15; (**c**) OC15; (**d**) ZHR100; (**e**) LP100; (**f**) EW100; (**g**) ZHR100 printed on the long edge; and (**h**) EW100 printed on the long edge. Sample LP100, printed on the long edge, will be shown later.

Sample LN100 (Figure 2a) shows a large area with stress whitening marks and, correspondingly, has a large residual strain after maximum deformation. GY15 (Figure 2b), oppositely, seems to have retained the original flat shape; however, in the right part of the sample, there are disrupted connections visible, being part of the full separation in the right half of the sample after the bending test. While samples OC15 (Figure 2c) and LP100 (Figure 2e) show residual deformations of the infill pattern, the other samples include broken strands along the lower plate of the sample where the pressure was applied.

Figure 3 shows the results of recovery tests during 10 cycles per sample, performed on the different specimens printed lying flat on the printing bed. Again, qualitative and quantitative differences between the infill patterns are visible.

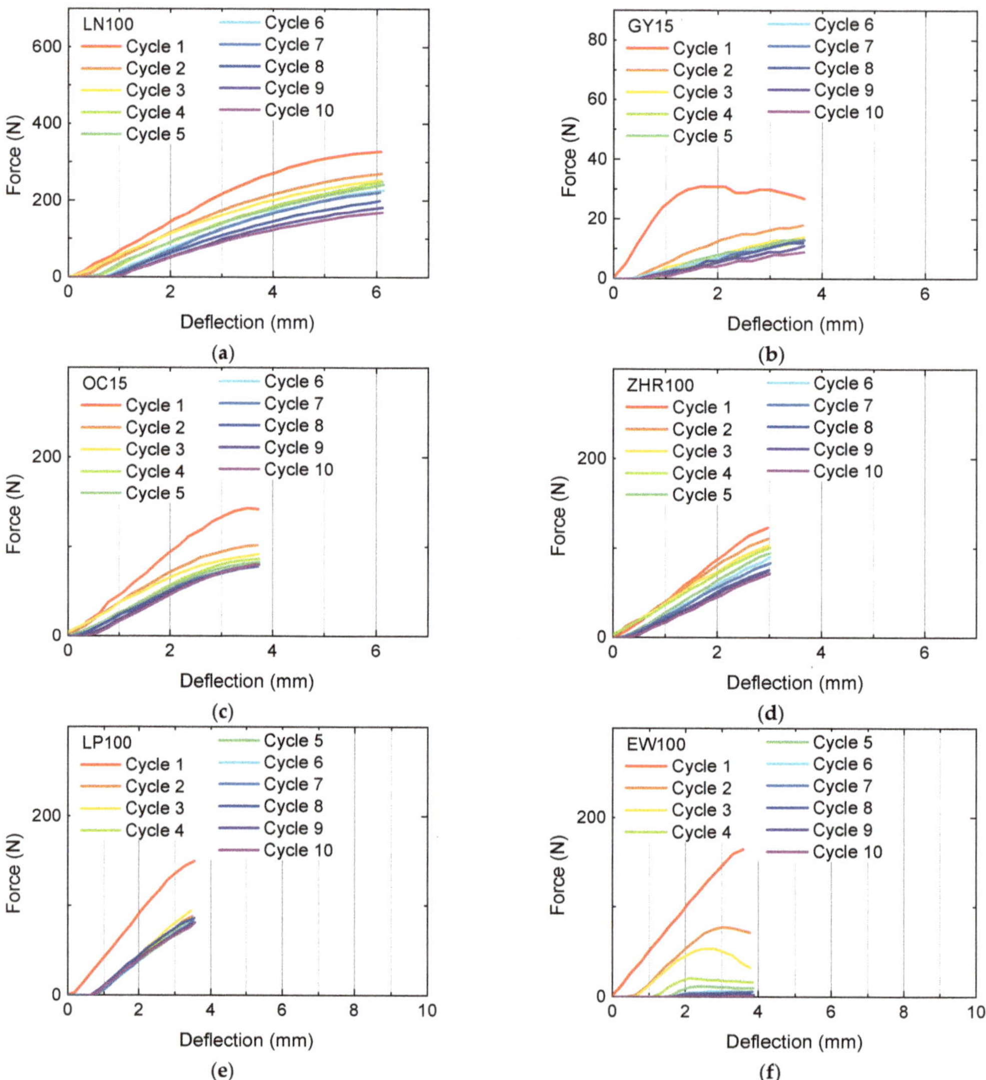

Figure 3. Bending and recovery tests of samples with different infill patterns: (**a**) LN100; (**b**) GY15; (**c**) OC15; (**d**) ZHR100; (**e**) LP100; and (**f**) EW100. Y-axis scales differ. Bending was performed till the deflection at which maximum force occurred, as measured in the previous tests.

For samples LN100 and ZHR100, the reduction in the maximum force after the first cycle is not much more pronounced than the consecutive reductions after the following cycles. Apparently, these infill patterns are better suited for bumpers, etc., which are normally damaged once or only a few times, as compared to the other samples showing a strong deviation between the first and the following cycles. The gyroid pattern shows a wave-like shape of the first force–deflection curve, similar to previous experiments with this pattern [15–17], and a relatively strong deviation between the first and second cycles, while the following cycles show relatively similar forces at identical deflections. However, none of the curves are smooth, which can be attributed to breaks of small connections occurring again and again.

The strongest deviations between the first and the second cycle are visible for sample LP100, while EW100 is nearly fully broken after 10 cycles, indicated by the very low forces measured at maximum deflection.

The corresponding microscopic images, taken after 10 bending and recovery cycles, are depicted in Figure 4. While all these samples show a slight residual strain, broken bonds are only visible in the lower part of EW100 (Figure 4f), indicating that the residual strain is correlated with changes in the material rather than with broken structures.

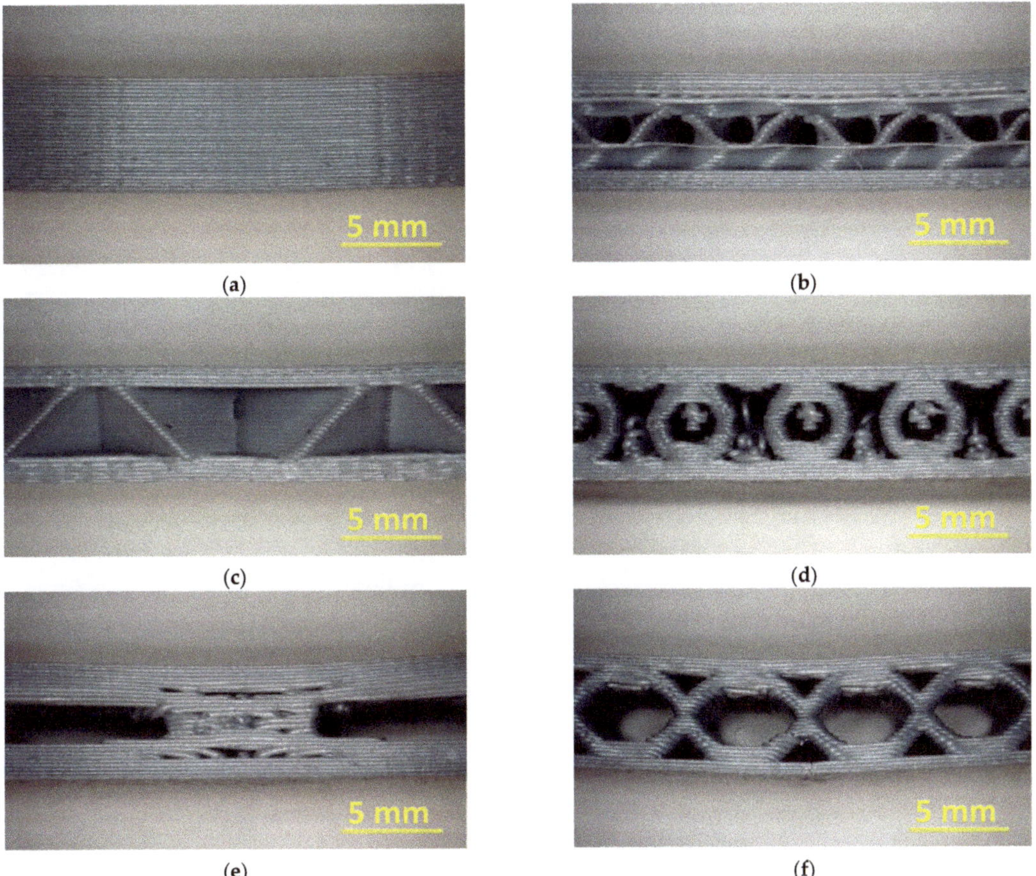

Figure 4. Microscopic images taken after 10 cycles of three-point bending and recovery tests of samples with different infill patterns: (**a**) LN100; (**b**) GY15; (**c**) OC15; (**d**) ZHR100; (**e**) LP100; and (**f**) EW100.

Next, Figure 5 shows the same tests, performed on sample LP100, printed on the edge. As mentioned before, due to the broad force maximum, different maximum deflections were chosen.

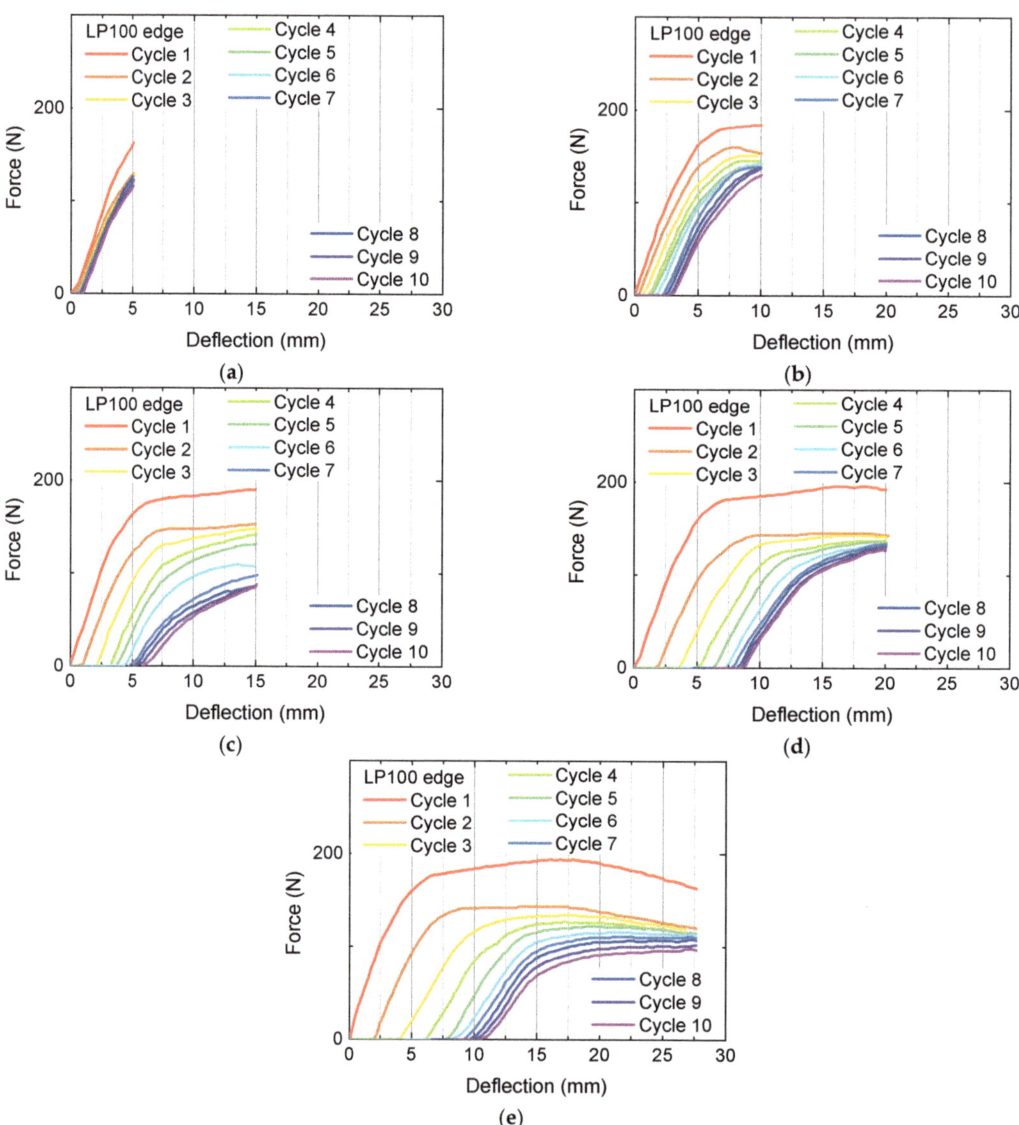

Figure 5. Bending and recovery tests of sample LP100, printed on the long edge, with different maximum deflections: (**a**) 5 mm; (**b**) 10 mm; (**c**) 15 mm; (**d**) 20 mm; and (**e**) 17.5 mm.

In all cases, the force differences between the first and the second curve are relatively large. In addition, the slopes of the curves differ not only for different cycles but also depending on the maximum deflection chosen. On the other hand, especially for the samples tested with a maximum deflection of 20 mm (Figure 5d), the forces reached at maximum deflection are quite similar for cycles 2–10, making this infill especially suitable for repeated deformations.

The corresponding microscopic images, taken after 10 bending and recovery tests each, are depicted in Figure 6. As expected, larger maximum deformations lead to larger residual strain. It should be mentioned that the full break visible in Figure 6c as well as the stress whitening observable in Figure 6d,e occurred usually after the first test cycle, after which no strong deviations were observed anymore, corresponding to Figure 5 showing that the largest force decrease occurs after the first cycle.

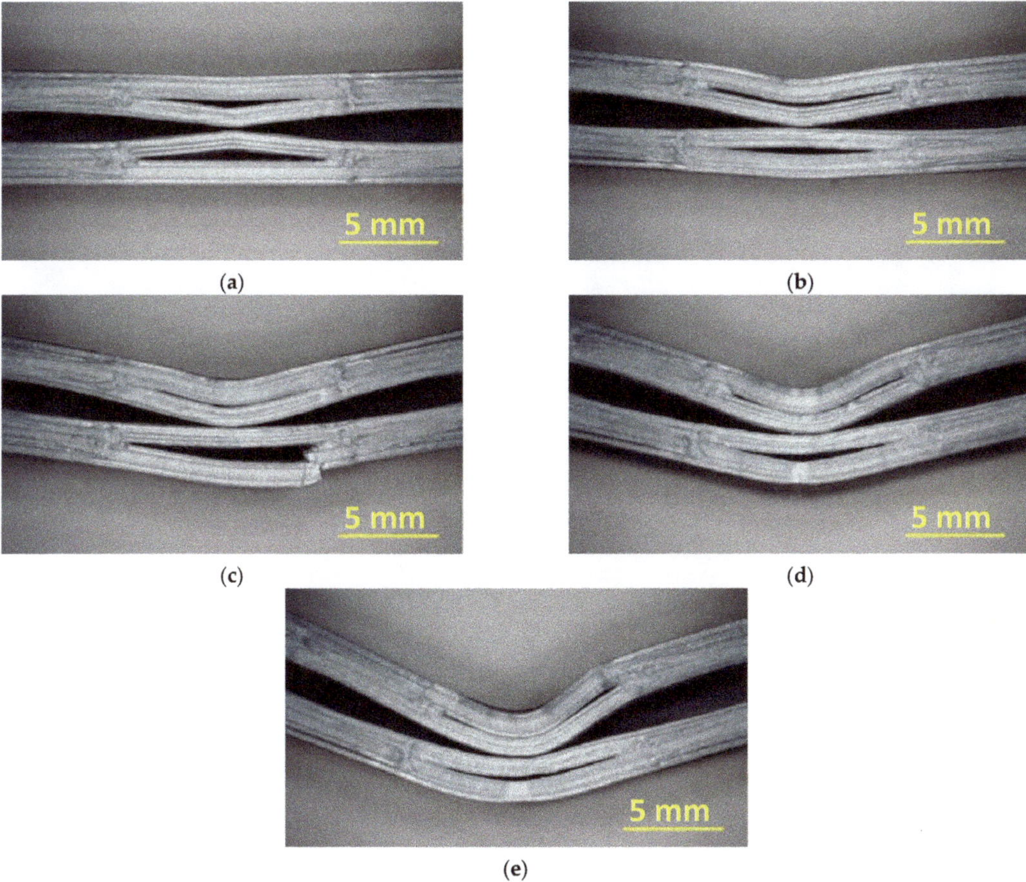

Figure 6. Microscopic images taken after 10 cycles of three-point bending and recovery tests of sample LP100, printed on the long edge, with different maximum deflections: (**a**) 5 mm; (**b**) 10 mm; (**c**) 15 mm; (**d**) 20 mm; and (**e**) 17.5 mm.

At the same time, these images suggest a possible optimization by varying the distance between the upper and lower halves of the leaf spring in the middle of the sample.

To compare the aforementioned residual strain quantitatively, Figure 7 depicts this value for all recovery tests shown here.

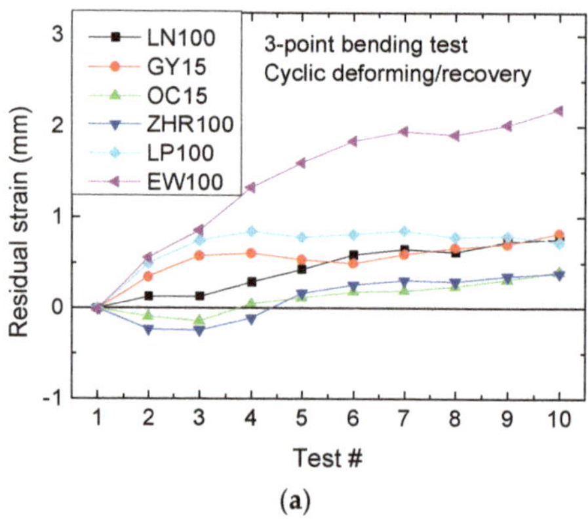
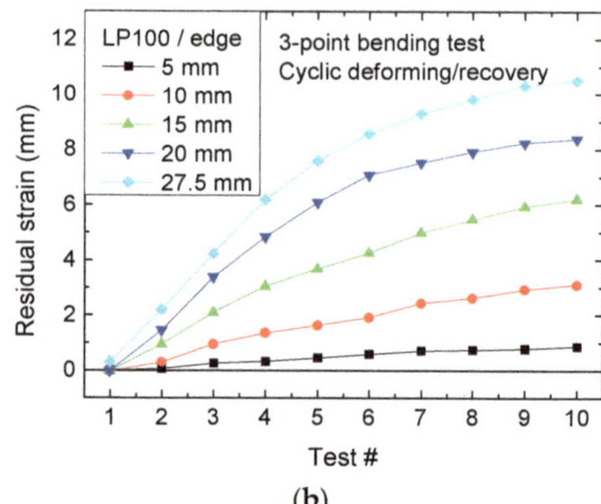

Figure 7. Residual strain of (**a**) samples printed flat; (**b**) sample LP100, printed on the edge, tested with different maximum deflections between 5 and 27.5 mm. Y-axis scales differ.

Amongst the samples printed flat on the printing bed (Figure 7a), EW100 shows the largest residual strain after 10 test cycles. For the infill patterns ZHR100 and OC15, the residual strain is firstly negative, meaning that the recovery process slightly over-compensates the deformation in the first three-point bending tests. This behavior was also found in previous tests, depending on the orientation of the impact [15–17].

For samples LP100, printed on the edge (Figure 4b), a clear correlation between the maximum deflection and residual strain is visible. Correspondingly, for the partly quite large deformation, residual strain values are reached which are much higher than those of the samples printed flat on the printing bed which were less bent.

Comparing these results with literature values is not easy since the test setups vary broadly. In many cases, researchers aim at the so-called 4D printing, meaning that deformation occurs above the glass transition temperature so that breaking of the samples is not to be expected [13,24–27]. Here, typically, recovery ratios near 100% are reached, if they are investigated at all; in many cases, one deformation is used to reach the desired final state.

In previous experiments of our group, cubes printed with different infill patterns and infill ratios were cold deformed along a linear impact area [15–17]. Here, the strong impact (with a depth of half the cube height) resulted generally in broken bonds which weakened the structure in subsequent test cycles. Similar to samples ZHR100 and OC15 (Figure 7a), a negative residual deformation was found in some cases, depending on the orientation of the applied pressure with respect to the printing orientation and thus to the infill pattern [17].

Senatov et al. compared pure PLA with PLA blended with hydroxyapatite (HA) to gain a certain self-healing effect by narrowing cracks after cold deformation. They found a shape recovery of 96–98% for the first cycle; however, delamination of the PLA sample already after the second cycle and of the PLA/HA sample after the third cycle, using porous structures, occurred [28]. Oppositely, here, most samples could be deformed without delamination or full break for the first ten test cycles. Nevertheless, it must be mentioned that the pressure was applied in a different way in Senatov's study, making it again hard to compare with our results. In another study, Senatov et al. found decreasing recovery stress values with increasing numbers of warm deformation (above the glass transition temperature) and recovery cycles, with PLA samples breaking after three cycles in this case, while PLA/HA samples could be recovered after warm deformation for a minimum of 10 cycles [29].

In the literature, three-point bending tests of PLA at low temperatures followed by recovery above the glass transition temperature are scarce. Liu et al. programmed samples from PLA and SiC/C/PLA into bent shapes or stretched them at a temperature of 90 °C and measured recovery at the same temperature, finding approximately 100% shape recovery [30]. Similarly, Dong et al. deformed three-point bending samples of PLA and PLA-grafted cellulose nanofibers above their glass transition temperature and found no difference between the recovery rates of these materials [31]. Liu et al. investigated a special textile-inspired structure in three-point bending tests; however, recovery tests were again performed after bending at higher temperatures [32].

As these examples show, research on the recovery of PLA after cold deformation is scarce and should be further extended to enable production of safety clothing, bumpers, etc., with shape recovery properties.

4. Conclusions

Three-point bending test samples were 3D printed from the shape-memory polymer PLA. Different infill patterns were chosen, some of which are typically given by common slicing software, while others were self-designed. For the latter, samples were printed lying flat on the printing bed or placed on the long edge. Besides testing samples up to maximum deflection, the specimens were subjected to a deflection until the maximum force was reached, recovered in a warm water bath, tested again, etc.

While the completely filled sample showed a smooth slope of the force–deflection curves and similar force reductions after each recovery cycle, samples with a gyroid or octet infill showed a stronger force reduction from the first to the second cycle, making these patterns less suitable for situations in which only few damages are expected.

Amongst the self-designed infill patterns, the leaf spring construction printed on the long edge showed, by far, the most interesting behavior. The very broad range of the approximate maximum force makes it suitable for security objects, such as bumpers, and allows recovery after different deflections. Here, nevertheless, it must be taken into account that the residual strain is increased with increasing maximum deflection.

These experiments show the strong impact of such infill patterns and, in some cases, of the printing orientations, suggesting further research to create structures with a similar broad range of the approximate maximum force to that in the leaf spring pattern, but, in addition, with a smaller residual strain, in order to make such structures technologically better usable.

Another aspect which will be taken into account in a future study is the impact of additives or fillers, as they are partly available in the so-called HT-PLA [23], and of a heat post-treatment which may both increase crystallinity to a certain amount [22] and thus not only the glass transition temperature but also the mechanical properties at room temperature.

Since investigations of recovery after low-temperature deformation are scarcely found in the literature, many more experiments are necessary to optimize structures and materials for 3D-printed SMPs for the potential use in bumpers, safety clothing and other objects which are deformed accidentally and could be more sustainable if they were to be recoverable at nearly 100%.

Author Contributions: Conceptualization, D.K. and A.E.; methodology, D.K. and A.E.; formal analysis, A.E.; investigation, D.K. and A.E.; writing—original draft preparation, A.E. and D.K.; writing—review and editing, D.K. and A.E.; visualization, D.K. and A.E. Both authors have read and agreed to the published version of the manuscript.

Funding: This research received no funding.

Institutional Review Board Statement: Not applicable.

Informed Consent Statement: Not applicable.

Data Availability Statement: The data presented in this study are fully available in this article.

Conflicts of Interest: The authors declare no conflict of interest.

References

1. Meng, H.; Hu, J.L. A brief review of stimuls-active polymers responsive to thermal, light, magnetic, electric, and water/solvent stimuli. *J. Intell. Mater. Syst. Struct.* **2010**, *21*, 859–885. [CrossRef]
2. Mather, P.T.; Luo, X.F.; Rousseau, I.A. Shape memory polymer research. *Annu. Rev. Mater. Res.* **2009**, *39*, 445–471. [CrossRef]
3. Blachowicz, T.; Pajak, K.; Recha, P.; Ehrmann, A. 3D printing for microsatellites—Material requirements and recent developments. *AIMS Mater. Sci.* **2020**, *7*, 926–938. [CrossRef]
4. Metcalfe, A.; Desfaits, A.-C.; Salazkin, I.; Yahiab, L.; Sokolowskic, W.M.; Raymonda, J. Cold hibernated elastic memory foams for endovascular interventions. *Biomaterials* **2003**, *24*, 491–497. [CrossRef]
5. Meng, Q.H.; Hu, J.L.; Yeung, L.Y. An electro-active shape memory fibre by incorporating multi-walled carbon nanotubes. *Smart Mater. Struct.* **2007**, *16*, 830–836. [CrossRef]
6. Lu, H.B.; Liu, Y.J.; Leng, J.S.; Du, S.Y. Qualitative separation of the physical swelling effect on the recovery behavior of shape memory polymer. *Europ. Polym. J.* **2010**, *46*, 1908–1914. [CrossRef]
7. Fu, C.-C.; Grimes, A.; Long, M.; Ferri, C.G.L.; Rich, B.D.; Thosh, S.; Ghosh, S.; Lee, L.P.; Gopinathan, A.; Khine, M. Tunable nanowrinkles on shape memory polymer sheets. *Adv. Mater.* **2009**, *21*, 4472–4476. [CrossRef]
8. Tobushi, H.; Hayashi, S.; Hoshio, K.; Miwa, N. Influence of strain-holding conditions on shape recovery and secondary-shape forming in polyurethane-shape memory polymer. *Smart Mater. Struct.* **2006**, *15*, 1033–1038. [CrossRef]
9. Chua, C.K.; Leong, K.F.; Lim, C.S. *Rapid Prototyping: Principles and Applications*, 2nd ed.; World Scientific Publishing Co. Pte. Ltd.: Singapore, 2003.
10. Wojtyla, S.; Klama, P.; Baran, T. Is 3D printing safe? Analysis of the thermal treatment of thermoplastics: ABS, PLA, PET, and nylon. *J. Occup. Environ. Hyg.* **2017**, *14*, D80–D85. [CrossRef]
11. Xu, J.; Song, J. Polylactic acid (PLA)-based shape-memory materials for biomedical applications. In *Shape Memory Polymers for Biomedical Applications*; Yahia, L.H., Ed.; Woodhead Publishing: Cambridge, UK, 2015; pp. 197–217.
12. Langford, T.; Mohammed, A.; Essa, K.; Elshaer, A.; Hassanin, H. 4D printing of origami structures for minimally invasive surgeries using functional scaffold. *Appl. Sci.* **2021**, *11*, 332. [CrossRef]
13. Mehrpouya, M.; Azizi, A.; Janbaz, S.; Gisario, A. Investigation on the functionality of thermoresponsive origami structures. *Adv. Eng. Mater.* **2020**, *22*, 2000296. [CrossRef]
14. Mehrpouya, M.; Gisario, A.; Azizi, A.; Barletta, M. Investigation on shape recovery of 3D printed honeycomb sandwich structure. *Polym. Adv. Technol.* **2020**, *31*, 3361–3365. [CrossRef]
15. Ehrmann, G.; Ehrmann, A. Shape-memory properties of 3D printed PLA structures. *Proceedings* **2021**, *69*, 6.
16. Ehrmann, G.; Ehrmann, A. Investigation of the shape-memory properties of 3D printed PLA structures with different infills. *Polymers* **2021**, *13*, 164. [CrossRef]
17. Ehrmann, G.; Ehrmann, A. Pressure orientation dependent recovery of 3D-printed PLA objects with varying infill degree. *Polymers* **2021**, *13*, 1275. [CrossRef]
18. Jia, S.K.; Yu, D.M.; Zhu, Y.; Wang, Z.; Chen, L.G.; Fu, L. Morphology, Crystallization and Thermal Behaviors of PLA-Based Composites: Wonderful Effects of Hybrid GO/PEG via Dynamic Impregnating. *Polymers* **2017**, *9*, 528. [CrossRef]
19. Wu, W.Z.; Ye, W.L.; Wu, Z.C.; Geng, P.; Wang, Y.L.; Zhao, J. Influence of layer thickness, raster angle, deformation temperature and recovery temperature on the shape-memory effect of 3D-printed polylactic acid samples. *Materials* **2017**, *10*, 970. [CrossRef]
20. Li, H.B.; Huneault, M.A. Effect of nucleation and plasticization on the crystallization of poly (lactic acid). *Polymer* **2007**, *48*, 6855–6866. [CrossRef]
21. Müller, A.J.; Ávila, M.; Saenz, G.; Salazar, J. Crystallization of PLA-based materials. In *Poly (Lactic acid) Science and Technology: Processing, Properties, Additives and Applications*; Jiménez, A., Peltzer, M., Ruseckaite, R., Eds.; RSC Polymer Chemistry Series No. 12; The Royal Society of Chemistry: Cambridge, UK, 2015.
22. Chalgham, A.; Wickenkamp, I.; Ehrmann, A. Mechanical properties of FDM printed PLA parts before and after thermal treatment. *Polymers* **2021**, *13*, 1239. [CrossRef]
23. Sölmann, S.; Rattenholl, A.; Blattner, H.; Ehrmann, G.; Gudermann, F.; Lütkemeyer, D.; Ehrmann, A. Mammalian cell adhesion on different 3D printed polymers with varying sterilization methods and acidic treatment. *AIMS Bioeng.* **2021**, *8*, 25–35.
24. Momeni, F.; Liu, X.; Ni, J. A review of 4D printing. *Mater. Des.* **2017**, *122*, 42–79. [CrossRef]
25. Kuang, X.; Roach, D.J.; Wu, J.T.; Hamel, C.M.; Ding, Z.; Wang, T.J.; Dunn, M.L.; Qi, H.J. Advances in 4d printing: Materials and applications. *Adv. Funct. Mater.* **2019**, *29*, 1805290. [CrossRef]
26. Sitotaw, D.B.; Ahrendt, D.; Kyosev, Y.; Kabish, A.K. Additive Manufacturing and Textiles—State-of-the-Art. *Appl. Sci.* **2020**, *10*, 5033. [CrossRef]
27. Koch, H.C.; Schmelzeisen, D.; Gries, T. 4D textiles made by additive manufacturing on pre-stressed textiles—An overview. *Actuators* **2021**, *10*, 31. [CrossRef]
28. Senatov, F.S.; Niaza, K.V.; Zadorozhnyy, M.Y.; Maksimkin, A.V.; Kaloshkin, S.D.; Estrin, Y.Z. Mechanical properties and shape memory effect of 3D printed PLA-based porous scaffolds. *J. Mech. Behav. Biomed. Mater.* **2016**, *57*, 139–148. [CrossRef]
29. Senatov, F.S.; Zadorozhnyy, M.Y.; Niaza, K.V.; Medvedev, V.V.; Kaloshkin, S.D.; Anisimova, N.Y.; Kiselevskiy, M.V.; Yang, K.-C. Shape memory effect in 3D printed scaffolds for self-fitting implants. *Europ. Polym. J.* **2017**, *93*, 222–231. [CrossRef]
30. Liu, W.B.; Wu, N.; Pochiraju, K. Shape recovery characteristics of SiC/C/PLA composite filaments and 3D printed parts. *Compos. Part A Appl. Sci. Manuf.* **2018**, *108*, 1–11. [CrossRef]

31. Dong, J.; Mei, C.T.; Han, J.Q.; Lee, S.Y.; Wu, Q.L. 3D printed poly(lactic acid) composites with grafted cellulose nanofibers: Effect of nanofiber and post-fabrication annealing treatment on composite flexural properties. *Add. Manufact.* **2019**, *28*, 621–628. [CrossRef]
32. Liu, Y.; Zhang, W.; Zhang, F.H.; Leng, J.S.; Pei, S.P.; Wang, L.Y.; Jia, X.Q.; Cotton, C.; Sun, B.Z.; Chou, T.-W. Microstructural design for enhanced shape memory behavior of 4D printed composites based on carbon nanotube/polylactic acid filament. *Compos. Sci. Technol.* **2019**, *181*, 107692. [CrossRef]

technologies

Article

Effect of Different Physical Cross-Linkers on Drug Release from Hydrogel Layers Coated on Magnetically Steerable 3D-Printed Microdevices

Roberto Bernasconi [1,*], Fabio Pizzetti [1], Arianna Rossetti [1], Riccardo Perugini [1], Anna Nova [1], Marinella Levi [2] and Filippo Rossi [1]

[1] Department of Chemistry, Materials and Chemical Engineering "Giulio Natta", Politecnico di Milano, Via Mancinelli 7, 20131 Milano, Italy; fabio.pizzetti@polimi.it (F.P.); arianna.rossetti@polimi.it (A.R.); riccardo.perugini@polimi.it (R.P.); anna.nova@mail.polimi.it (A.N.); filippo.rossi@polimi.it (F.R.)

[2] Department of Chemistry, Materials and Chemical Engineering"Giulio Natta", Politecnico di Milano, Piazza Leonardo da Vinci 32, 20133 Milano, Italy; marinella.levi@polimi.it

* Correspondence: roberto.bernasconi@polimi.it; Tel.: +39-022-399-3150

Abstract: In the last few decades, the introduction of microrobotics has drastically changed the way medicine will be approached in the future. The development of untethered steerable microdevices able to operate in vivo inside the human body allows a high localization of the therapeutical action, thus limiting invasiveness and possible medical complications. This approach results are particularly useful in drug delivery, where it is highly beneficial to administer the drug of choice exclusively to the target organ to avoid overdosage and side effects. In this context, drug releasing layers can be loaded on magnetically moveable platforms that can be guided toward the target organ to perform highly targeted release. In the present paper, we evaluate the possible application of alginate hydrogel layers on moveable platforms manufactured by coupling additive manufacturing with wet metallization. Such alginate layers are reticulated using three different physical crosslinkers: Ca, Zn or Mn. Their effect on drug release kinetics and on device functionality is evaluated. In the case of alginate reticulated using Mn, the strongly pH dependent behavior of the resulting hydrogel is evaluated as a possible way to introduce a triggered release functionality on the devices.

Keywords: microrobots; 3D printed; drug delivery; hydrogels; alginate

Citation: Bernasconi, R.; Pizzetti, F.; Rossetti, A.; Perugini, R.; Nova, A.; Levi, M.; Rossi, F. Effect of Different Physical Cross-Linkers on Drug Release from Hydrogel Layers Coated on Magnetically Steerable 3D-Printed Microdevices. *Technologies* **2021**, *9*, 43. https://doi.org/10.3390/technologies9020043

Academic Editors: Nam-Trung Nguyen and Dennis Douroumis

Received: 2 May 2021
Accepted: 11 June 2021
Published: 18 June 2021

1. Introduction

Since ancient times, humankind has discovered that specific substances, both available in nature and synthetic, can be used to cure or even prevent a large number of pathologies. Starting from a primordial knowledge based on herbalism and natural ingredients [1], pharmacology has evolved in modern times into a systematic and technologically advanced science. However, while the research on active principles has significantly progressed, the quest for new administration routes has not proceeded in a comparable way. Even nowadays, the majority of medical preparations follow pharmacokinetics still based on poorly controllable distribution routes in the body. Enteral, intravenous and intramuscular administration all rely on blood-mediated distribution of the drug, while transdermal delivery relies on diffusion through the skin [2]. Indiscriminate transport in the whole body translates into higher doses required to reach a therapeutic concentration in the target organ, with possible dosage-related counterindications [3]. Moreover, non-optimal drug usage may possibly induce drug resistance in the case of pathogens [4] or cancer cells [5].

Starting from these premises, it is evident that targeting drug delivery is a matter of major importance in modern pharmacology [6,7]. In the last few decades, a wealth of smart delivery approaches has been proposed at the laboratory scale: liposomes [8,9], polymeric thin layers [10], DNA nanostructures [11], dendrimers [12], micelles [13] and biocompatible nanoparticles [14–16]. All these approaches partially addressed the problem of temporal control over release, but did not significantly address the problem of carrying the drug directly in correspondence of the target organ. One of the most promising approaches to carry out this task is to load the drug-bearing material on remotely guided microrobots able to navigate the body in vivo and reach the target organ [17,18]. Such devices have been developed in the last few years and their potential efficiency for targeted drug administration has been widely demonstrated [19,20]. Biomedical microrobots are, in the vast majority of the cases, actuated by applying controlled, non-invasive magnetic fields [21]. They can be manufactured in a wide dimensional range, from millimeters [22,23] to a few micrometers [24,25], according to the characteristic dimensions of the target organ (e.g., centimeters for the gastrointestinal apparatus or micrometers for some blood vessels).

From the fabrication point of view, magnetically guidable microdevices can be conveniently fabricated using many different techniques. The most interesting, in terms of customizability and flexibility, is probably 3D printing [26]. Drug releasing devices can be printed directly using a drug loadable soft polymer [25,27,28] or using rigid materials that are subsequently coated with drug-releasing polymers [23,29]. In both cases, hydrogels are between the most interesting materials usable [30]. These natural macromolecules are well-known for their biocompatibility, efficiency in drug loading and ease of manufacturing. Furthermore, they can be opportunely functionalized to release drugs only in well-specified conditions (of pH, temperature, etc.) [31]. In this context, we recently demonstrated that untethered microdevices can be fabricated by coupling 3D printing with wet metallization [22] and that the same technique can be used to realize hydrogel-coated microrobots for controlled drug release [32].

In the present work, we explored the effect of biocompatible crosslinkers alternative to calcium chloride on the drug release properties of alginate layers coated on 3D-printed untethered microdevices. Following the research trend recently started by our group [22,23,33], we employed 3D printing in the form of stereolithography to efficiently manufacture miniaturized moveable platforms that we subsequently coated with different functional metallic layers. In detail, microrobots were coated with CoNiP to allow magnetic actuation and with gold to provide a biocompatible surface. The moveable platforms obtained in this way were then coated with alginate hydrogels crosslinked with either $ZnCl_2$ or $MnCl_2$. These two selected alternative crosslinkers have recently been investigated by Da Silva et al. [34], who did not verify their drug release properties. Consequently, the first aim of the present work is to verify the applicability of alginate layers on shape-optimized moveable platforms and to evidence differences in drug delivery properties. The second main aim of the work is to investigate alternative methodologies to trigger release from alginate exploiting pH variations. In our previous paper [32], we functionalized alginate hydrogels by introducing a pH cleavable bond that allowed drug release only in a well-defined pH range. By doing this, we were able to trigger drug release according to the pH of the environment crossed by the device. The main drawback of the approach is represented by the necessity to carry out a chemical synthesis, which can potentially damage the molecule loaded and must be specifically designed for each drug. Da Silva et al. evidenced that Mn-reticulated alginate mechanically degraded at low pH. This particularity was employed to implement a pH-governed instantaneous release functionality on the microdevices. From the applicative point of view, microdevices carrying pH-sensitive hydrogels can find potential application in the gastrointestinal tract, which is characterized by significant pH variations [35].

2. Experimental Methods

2.1. Microdevices 3D Printing

The miniaturized devices employed in the present work were 3D-printed and metallized following our previous work [33]. Concisely, the geometry of the devices was initially optimized and their 3D model was designed using Solidworks (Dassault Systèmes, France). The geometry was optimized for stereolithography using Nauta+ (DWS, Thiene, Italy) and the 3D model was sliced with Fictor (DWS, Thiene, Italy). Micro-stereolithography was carried out using a model 028 J Plus setup (DWS, Thiene, Italy) and the resulting prints were post-cured by exposition to UV radiation for 30 min. The 028 J Plus setup comprises a galvanometer control and a laser able to yield a power of 30 mW at a wavelength of 405 nm, with a beam spot diameter of 22 µm.

2.2. Microdevices Metallization

A detailed description of the metallization process can be retrieved in our previous work [33]. Briefly, the printed devices were removed from printing supports and coated with a first layer of copper to make their surface conductive. Subsequently, they were coated with CoNiP exploiting a barrel-like metallization approach. At the end of the CoNiP metallization step, the procedure was varied with respect to the cited literature reference. In place of Cu and Ag/TiO_2, the samples were coated (always via barrel-like plating) with a gold layer from electrolytic deposition in a cyanide-free electrolyte (SG-Au 340 bath from SG Galvanobedarf GmbH). The following parameters were employed: $5 \text{ mA}/cm^2$, moderate stirring, 50 °C. At the end of the gold plating step, the samples were thoroughly washed with deionized water and dried with nitrogen.

2.3. Hydrogel Application

To apply the hydrogel layers on the devices, the latter were dipped sequentially into the polymer aqueous solutions and then in the physical crosslinker in order to deposit the layers (Figure S7). With the goal of obtaining a homogeneous coating of the device, a custom-made structure (Figure S8) was utilized. The device was suspended to a very thin nylon wire, preventing it from touching any part of the beakers containing the solutions. Initially, the device was immersed into a 2% *m/v* pure sodium alginate solution (step a in Figure S7). This step was performed inside an ultrasonic bath in order to favor the penetration of the alginate in the scaffold. Then, the device was removed from the alginate solution and immersed into a second solution containing the physical cross-linker (step b in Figure S7). Following the work by Da Silva et al. [34], three physical crosslinkers were employed: $MnCl_2$ (1% *w/v* in DI), $ZnCl_2$ (1% *w/v* in DI) and $CaCl_2$ (1% *w/v* in DI).

2.4. Microdevices Characterization

SEM was performed by means of a Zeiss EVO 50 setup, suitably equipped with an Oxford Instruments Model 7060 EDS module. The magnetic properties of the devices were determined using of a Princeton Measurement Corp. MicroMag 3900 vibrating sample magnetometer (VSM). The roughness of the deposited gold layer was evaluated using a UBM Microfocus laser profilometer.

2.5. In Vitro Drug Delivery

Drug release was investigated in simulated physiological conditions: at 37 °C and 5% CO_2, in a phosphate-buffered saline solution (PBS, pH 7.4). In detail, each device was placed in excess of PBS (2.5 mL) and aliquots were collected at defined time points, replacing them with an equal volume of fresh solution in order to preserve the diffusion regime between the device and the release environment. Percentages of released Rhodamine B (RhB) were then measured by UV spectroscopy at 570 nm.

2.6. Magnetic Actuation

Devices were remotely actuated employing the Octomag magnetic manipulation setup [36]. Microrobots were permanently magnetized along the direction perpendicular to their axis by placing them on a strong NdFeB magnet. Then, a rotating magnetic field characterized by varying intensity and frequency (τ) was applied thanks to the Octomag. Uncoated and hydrogel-coated devices were actuated inside a water-filled glass basin to avoid hydrogel desiccation. Their motion was tracked and their speed determined using the software Tracker.

3. Results and Discussion

3.1. Shape Optimization and 3D Printing

Recently, we described the use of magnetically moveable scaffold-like architectures to support and transport pH-sensitive hydrogels [32]. The choice of using a scaffold-like design was motivated by the capability of a porous structure to host a good quantity of hydrogel inside its pores. Retrospectively, however, more optimized shapes were evaluated to increase the quantity of hydrogel loaded. The inspiration for a more suitable design was taken from the so-called honey dipper (Figure S1), a common household device normally employed to dose highly viscous fluids (e.g., honey). The particular shape of this tool, characterized by the presence of different parallel plates, is optimized to efficiently retain the fluid in which it is immersed. Since the alginate solution is a viscous fluid prior to gelation, the same concept can be transferred to the design of untethered microdevices. Millimetric honey dippers, characterized by the dimensions detailed in Figure S2, were designed and printed using stereolithography. In analogy with our previous work [32], dimensional features were kept in the few mm to hundreds μm dimensional range. This is a consequence of the potential use of the devices in the gastrointestinal apparatus, whose tracts present characteristic dimensions compatible with these microdevices. A typical batch of devices was composed of 10 to 15 units and it was successfully printed in a roughly 30 min timespan. Devices were printed exploiting small supports connected to individual printing bases (Figure 1a). At the end of the 3D printing step, devices were removed from the supports using a cutter (Figure 1b).

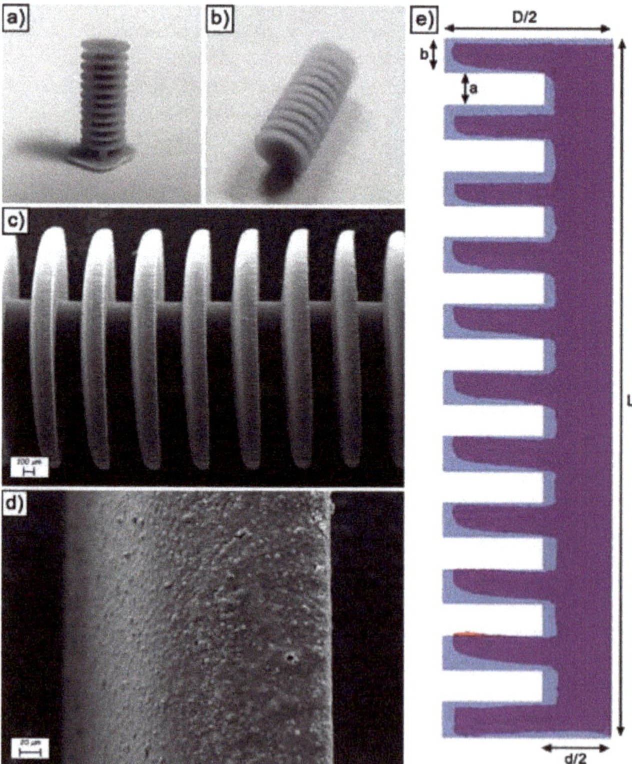

Figure 1. Visual appearance of an as printed device before (**a**) and after (**b**) printing supports removal; SEM images of an as printed device observed at 100 X (**c**) and 1000 X (**d**); visual comparison between the experimental and theoretical section of a device (**e**).

As-printed devices were characterized from the morphological and dimensional point of view. Figure 1c is a low-magnification SEM image, in which the general structure of the devices is clearly visible. As expectable from a micrometric SLA printing, the edges are considerably rounded due to the comparatively large laser beam size (22 μm). By increasing the magnification (Figure 1d), the microstructure of the SLA resin is evident. In particular, it presents a relatively rough surface due to the presence of a silica/aluminate filler inside. The latter is clearly visible in the form of spherical particles by further increasing the magnification (Figure S3).

To evaluate dimensional conformity to the original 3D model, devices were sectioned and inglobated inside an epoxy resin. The surface was then polished, yielding the result visible in Figure S4. The profile of the section was then extracted and superimposed to the theoretical one (Figure 1e). In general, the experimental profile is relatively adherent to the 3D model. The comparison, however, confirms the edge rounding observed at the SEM and allows to quantify the deviation from the ideality (Table S1). Large features present experimental dimensions relatively adherent to the corresponding theoretical values (98.9% for L, 95.8% for D and 84.6% for d). The largest deviations from the ideality were observed on the stacked disks that constitute the body of the device, which are also the smallest features of the design and present the most challenging printing conditions. In detail, the mean thickness of the disks (b) is only 61.3% of the theoretical value, while the distance between them (a) is 140.5% of the theoretical value. Both these dimensions also present a remarkable variability.

3.2. Devices Metallization

After printing, the devices were metallized. Unlike our previous paper on hydrogel-coated microdevices [32], the different metallic layers were not entirely applied by means of electroless deposition. Electrolytic deposition was used instead, following another previous paper published by our group [33]. By doing this, metallic layers characterized by more controllable properties and thickness were deposited. The first layer, due to the non-conductive nature of the SLA resin, was obligatorily applied by means of electroless deposition. Consequently, 400 nm of Cu were applied via immersion for 15 min in a formaldehyde-free copper electroless bath. Then, once the surface was conductive, the barrel-plating approach was employed to apply CoNiP and gold from electrolytic plating. This technology, which takes inspiration from the industrial practice used to coat small objects and is extensively described in our previous publication [33], allows electrolytic deposition without directly contacting each device. In turn, they are placed inside an electrified metallic basket and plated all together, exploiting the labile electrical connection between the devices and the walls of the basket (Figure 2a). A total of 5 μm of CoNiP, whose ferromagnetic properties were exploited to make the devices actuable via magnetic fields, were deposited on the microrobots in 160 min.

With regard to the barrel plating approach, we observed in our previous work [33] that the non-ideal contact between the devices and the basket requires the introduction of a correction coefficient (named ψ) in the standard well-known Faraday law Equation (1). The relationship describes the amount of material deposited with respect to the total charge used.

$$m = \frac{Mq}{ZF} \eta \, \psi \tag{1}$$

m is the total mass reduced, M is the molar mass, q is the total charge, Z is the valence of the ions reduced, F is the Faraday constant, η is the cathodic efficiency of the reaction. ψ, as previously stated, is an apparent efficiency of the barrel process, which must be determined by comparing the amount of metal deposited in the barrel and the corresponding amount plated in analogous conditions on a standard planar surface with ideal electrical connection to the generator. η and ψ for CoNiP deposition were determined previously [33] and they resulted equal to 0.42 and 0.25, respectively. In other words, only 10.5% of the current in the barrel is used to actually plate CoNiP on the devices. However, this apparent disadvantage of the barrel-plating approach is counterbalanced by the possibility to plate layers on small objects using electrolytic deposition and without leaving uncoated areas due to electrical contacting.

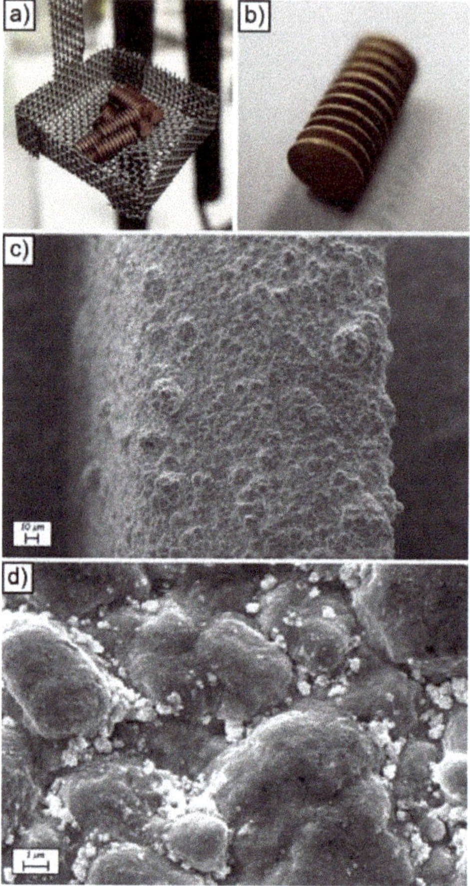

Figure 2. Visual appearance of the barrel setup used to metallize the devices (**a**); visual appearance of a Cu/CoNiP/Au coated device (**b**); SEM images of a metallized device observed at 1000 X (**c**) and 20,000 X (**d**).

At the end of the CoNiP plating step, the microdevices were coated with a 4-µm thick gold top layer to make the surface biocompatible and to avoid corrosion. Such a layer was again deposited using the barrel approach in a non-cyanide gold-plating bath. With respect to the galvanic displacement methodology previously employed to apply gold [22], electrolytic deposition allows to deposit thick layers (in the few µm range). Increasing the gold thickness renders the devices more resistant to corrosion, thus increasing the number of reuse cycles. Barrel gold deposition was carried out here for the first time and, for this reason, η and ψ were determined. η of the deposition process was evaluated depositing gold on a planar copper substrate and the result was equal to 0.57. ψ was evaluated by weighing the devices before and after deposition in the barrel, considering the cathodic efficiency previously calculated and comparing the result with the mass deposited on the planar copper sample. The correction coefficient ψ resulted equal to 0.28. Gold was deposited for 220 min to obtain the thickness required.

The final visual appearance of a gold-coated device is visible in Figure 2b. The surface appears uniformly covered with gold, as further certified by the SEM analysis (Figure 2c,d and Figure S5) and by the EDS characterization (Figure S6). As expected, the application of a metallic layer on the surface of the devices increased the roughness of the surface from 191 ± 23 nm (R_a for the surface of the uncoated resin) to 356 ± 55 nm (R_a for the surface $Cu/CoNiP/Au$ coated devices). SEM analysis evidenced a nodular morphology on the surface of the devices, which was a result of the electrolytic deposition process employed for their metallization.

3.3. Hydrogels Application

At the end of the metallization process, microdevices were coated with alginate and the effect of the three different physical cross-linkers was examined. Considering the results obtained in different papers [37], the concentration of sodium alginate was fixed at 2% w/v in distilled water. Too high alginate concentrations lead to stiff hydrogels, which show higher rigidity but lower swelling abilities, thus compromising the drug loading capacity and its utilization inside the human body. Furthermore, the solutions with higher concentration are too viscous with a very long time of complete dissolution, and do not facilitate the deposition of the hydrogel on the device. Indeed, by increasing the alginate concentration, there are more polar groups (-COO-) interacting with divalent ions of the cross-linker agent, increasing the cross-linking density [38]. The crosslinking process keeps the chains closer to each other, increasing the mechanical properties of the devices but reducing their water and drug uptake. Lower concentrations produce, instead, a hydrogel with an excessive swelling degree, thus decreasing the adhesion to the substrate, increasing the volume and having inconsistent shapes with the magnetic actuation. Therefore, the final choice resulted from a balance between the swelling degree and the adhesion and consistency of the hydrogel over the scaffold.

In the present work, only the type of cross-linker was varied to evidence its effect on release properties. In general, alginate has the ability to form a 3D hydrogel network when in contact with divalent cations (such as Ca^{2+}, Fe^{2+}, Mn^{2+}, Ba^{2+}, Sr^{2+}). Here, in analogy with Da Silva et al. [34], three different biocompatible cross-linker agents have been tested: $MnCl_2$, $ZnCl_2$ and $CaCl_2$. The relative stoichiometry of the crosslinker was kept constant at a value of 1% w/v, which has been reported in many works as optimal from the swelling, uniformity and stability point of view [34,37,38]. Indeed, hydrogels should remain attached over the scaffolds, without collapsing, with a thickness as uniform as possible, and they must permit a good uptake of water to guarantee a sufficient loading of drug molecules. Low cross-linker concentration means a lower diffusion gradient that leads to low rates and low degrees of crosslinking. This implies a very high swelling degree, excessive volume and poor homogeneity. On the contrary, excessive cross-linking can reduce the soft nature of the system needed for applications with living tissue. Lastly, higher concentrations of the crosslinking agent favor its diffusion among alginate chains and determine a faster crosslinking process.

Devices were suspended to a thin wire and sequentially immersed in the alginate solution and then in the crosslinking solution (Figure S7). To sustain the wire, a self-designed holder (Figure S8) was employed. In line with our previous work [32], the total thickness of the hydrogel layer could be increased by repeating the coating sequence. However, as evidenced in Figure 3a–d for the Ca^{2+} reticulated alginate, the application of more than one layer progressively altered the shape of the devices. With one layer (Figure 3a) and two layers (Figure 3b), the shape of the devices remained relatively cylindrical. At three (Figure 3c) and four (Figure 3d) layers, the final shapes of the devices were strongly rounded by the surface tension of the non-reticulated hydrogel. The ellipsoidal shape resulting from more than two coating sequences could potentially interfere with device actuation, which was designed for cylindrical-shaped devices. For this reason, the number of hydrogel layers was limited to a maximum of two, ideally one. With this consideration, it is fundamental to load the highest possible amount of hydrogel to maximize the amount of loadable drug.

The shape of the device has a strong influence on this aspect, as demonstrated by calculating the weight difference between uncoated and single layer coated devices. Hydrogel uncoated devices were characterized by a weight of 31.46 ± 1.78 mg. Considering Ca^{2+} as a crosslinker, the final weight of the loaded hydrogel was 31.5 ± 5.5 mg. This translated into a 100.13% increase of the weight after coating. If compared with the increase registered in our previous work [32] with porous microdevices (80%), this value demonstrates that the honey dipper design is able to optimize hydrogel loading after a single coating step. Besides the device shape, the crosslinker employed also influenced the final loaded weight. Indeed, Figure 3e depicts the weight increase recorded in the case of alginate applied on the microdevices and reticulated with Ca, Zn and Mn. A clear influence of the reticulating agent can be observed, with Mn allowing the lowest amount of loaded hydrogel (14 ± 0.5 mg).

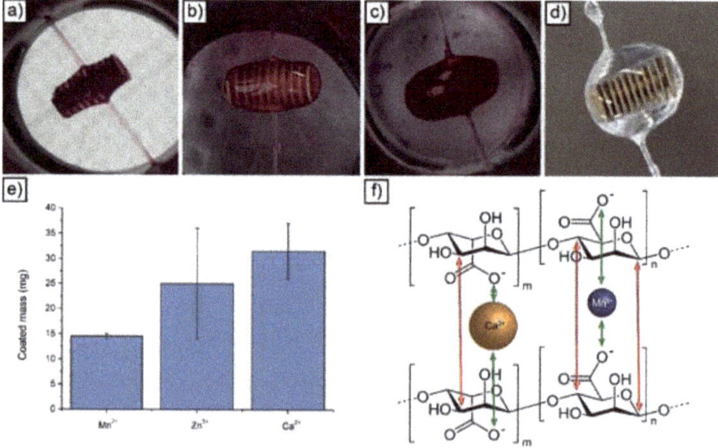

Figure 3. Visual appearance of devices coated with Ca-reticulated hydrogel after one (**a**), two (**b**), three (**c**) and four (**d**) coating cycles; amount of applied hydrogel after a single coating cycle as a function of the physical crosslinker employed (**e**); schematic representation of two alginate chain crosslinked with either Ca or Mn (**f**).

From the results obtained by Da Silva et al., it was possible to see that the hydrogel prepared with Mn^{2+} had a higher swelling degree if compared to those prepared with Zn^{2+} or Ca^{2+}. As the three cross-linkers have the same valence ($^{2+}$), the swelling variation can only be attributed to the different size of the ions—0.067, 0.074 and 0.100 nm for Mn, Zn and Ca, respectively. Ions with a larger radius have a higher interaction with the groups -COO- of sodium alginate, thus generating chains with higher entanglement (Figure 3f). The stronger interactions of the larger ions with the polymeric chain may be related to the polarizability and London dispersion forces. Larger molecules, atoms or ions usually exhibit higher London dispersion forces and are also more polarizable, thus increasing ion-dipole interactions with the polymer chain. These interactions minimize the elasticity of the polymeric chain, reducing its swelling degree. In addition, as is evident from Figure 3f, larger ions occupy a larger space in the empty interstices among the polymer chains (or pores), decreasing the virtual volume which could be occupied by water molecules. Therefore, due to their size, calcium ions are more efficient (in the same volume) in the polymeric chain entanglement, also providing more intermolecular interactions and limiting the expansion of the polymeric chain [34]. On the contrary, Mn^{2+} ions, due to the lowest ionic radius compared to the others, present a weak interaction with the -COO- groups of the alginate. This fact limits the effective crosslinking but improves the water absorption rate. This leads to less rigid hydrogels with polymer chains relatively free to move.

From these considerations, it can be inferred that for the affinity of the ions with -COO- groups, the best crosslinkers, in order, are: $Ca^{2+} > Zn^{2+} > Mn^{2+}$. From the experiments, it was also noticed that the mass of hydrogel loaded over the scaffold seems dependent on the type of crosslinker agent used. In particular, it also seems that the amount of gel loaded on the scaffold follows the same scale of affinity. As shown in Figure 3e, Ca^{2+} cations, with the same procedure, allow to load more gel over the scaffold with respect to the two other ions, all without losing good homogeneity and good stability properties.

Another aspect to take into consideration is the effect of pH on the swelling behavior. In agreement with the results obtained by Da Silva et al. for the hydrogels crosslinked with Mn^{2+}, it does not seem possible to measure the water absorption at different pH levels since the hydrogels dissolved during the first hours of the study. This result is interesting because it shows that the physical interactions generated in the synthesis of the hydrogels (using Mn^{2+} as crosslinking agent) can be easily broken. This thermodynamic instability was attributed to the competition between the H^+ ions of the swelling medium and the Mn^{2+} ions. In particular, the carboxylic groups responsible of the interactions with the divalent ions Mn^{2+} in the crosslinking process started to interact with the monovalent cationic ions of the medium (Na^+, H^+), in the form of carboxylic acid or carboxylic acid salt ($-COOH$, $-COONa$) groups, destroying the three-dimensional network. As to the effect of pH on the hydrogels crosslinked with Zn^{2+} or Ca^{2+}, just a change in the swelling behavior, instead of a complete dissolution, was verified. This is due to the interactions of carboxylic groups with these two cations that are stronger and therefore only partially altered by Na^+ or H^+ action.

3.4. Drug Release Performances

Considering the different swelling behavior observed by Da Silva et al. [34] for the three different crosslinkers, an observable difference was also expected in the case of drug release performances. RhB release was investigated and the behavior of the three hydrogels was compared. Release studies were conducted at 37 °C and pH 7.4, with the results visible in Figure 4a. The percentage of RhB released was defined as the ratio between the released amount in the aqueous media and the total amount loaded within the polymeric layer.

It appears evident that, contrary to what expected, the three materials followed a comparable trend, which led to an almost complete release of RhB after roughly two hours. This is in accordance with the fact that RhB presents a very small hydrodynamic radius and its release is not influenced by the different cross-linkers used. Information about the kinetics of the release can be extracted from the data visible in Figure 4a if the same are plotted against the time square root (Figure 4b). The initial linear part of the release curve is indicative of Fickian diffusion and the y-axis intercept value is an indication of burst release. The latter is relatively limited and took place when RhB-loaded devices were placed in the releasing medium. After the initial fast release, the RhB loaded within the three alginate hydrogels showed a marked linear trend only in the first 2 h. This corresponds to a pure Fickian diffusion and is only driven by the concentration gradient. Afterwards, the trend reached a plateau. Indeed, when the main release was completed, the residual drug entrapped within the hydrogel network was slowly released.

The data obtained were used to estimate RhB diffusion coefficients. Operatively, the release mechanism could be considered as a pure Fickian diffusion, the concentration being driven through alginate. Under this assumption, the drug diffusion kinetic can be described as a one-dimensional model of the second Fick law where the device geometry is a cylinder and the material flux mainly takes place at the PBS/hydrogel surface. Equation (2) showed these considerations, indicating r as the characteristic radius for the mass transport phenomenon. The following mass balance equations are written considering the variation of the mean drug concentration within the hydrogel (C_G) related to the volume of solution (V_S), the mean drug concentration in the outer solution (C_S), the total volume (V_G), the drug present inside the matrix (m_G) and the exchange interfacial surface (S_{exc}), which represents the boundary surface between the device and the surrounding solution (which, simplifying, can here be considered as being only the side surface). According to these expressions, the boundary conditions are defined describing the profile symmetry at the center of the polymeric cylinder with respect to the radial axis of the cylinder—Equation (6)—and the equivalence between the material diffusive fluxes at the PBS/hydrogel surface—Equation (7).

$$\frac{\partial C_G}{\partial t} = D \cdot \frac{1}{r^2} \cdot \frac{\partial}{\partial r} \left(r^2 \cdot \frac{\partial C_G}{\partial r} \right) \tag{2}$$

$$V_S \frac{\partial C_G}{\partial t} = k_C \cdot S_{exc} \cdot (C_G - C_S) \tag{3}$$

$$C_S(t = 0) = 0 \tag{4}$$

$$C_G(t = 0) = C_{G,0} = \frac{m_{G,0}}{V_G} \tag{5}$$

$$\left. \frac{\partial C_G}{\partial r} \right|_{r=0} = 0 \tag{6}$$

$$-D \cdot \left. \frac{\partial C_G}{\partial r} \right|_{r=R} = k_C \cdot (C_G - C_S) \tag{7}$$

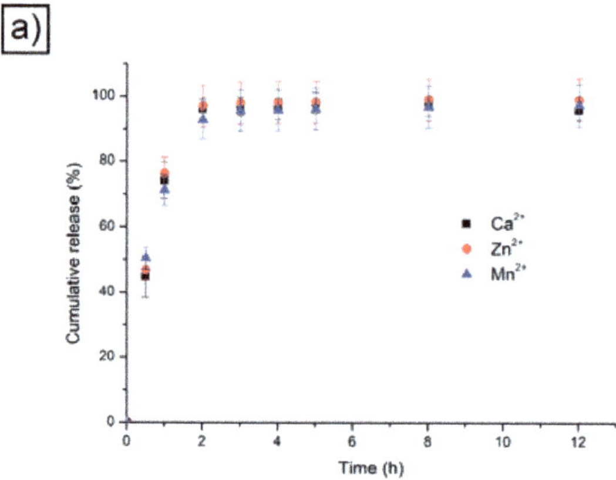

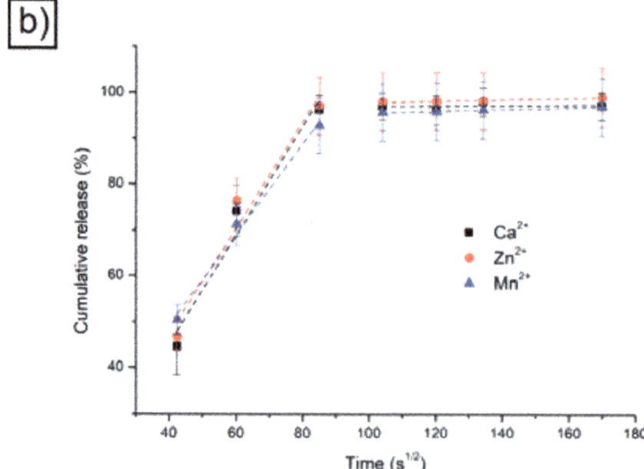

Figure 4. RhB release results obtained at 37 °C and pH 7.4 with different physical crosslinkers (**a**); RhB release results obtained at 37 °C and pH 7.4 with different physical crosslinkers plotted against time square root (**b**).

This mathematical model allowed to estimate the diffusion coefficient (D) of RhB. The Sherwood number obtained by means of penetration theory allowed the computation of the mass transfer coefficient k_C—Equation (8).

$$Sh = 1 = \frac{k_C \cdot 2r}{D} \tag{8}$$

Table 1 resumes the results obtained.

Table 1. Diffusion coefficient of RhB in alginate hydrogels reticulated with different cross-linkers.

Title 1	Diffusivity [cm^2/s]
Ca reticulated alginate	2.56 10^{-4}
Zn reticulated alginate	2.7 10^{-4}
Mn reticulated alginate	2.57 10^{-4}

As expected from the data reported in Figure 4, the three diffusion coefficients are comparable. This result confirms the limited effect induced by the difference in ionic size of the three crosslinkers on the Rhb release.

Apparently, by considering the results obtained, the most advantageous crosslinking agent for alginate is calcium. Indeed, devices can be loaded with a higher amount of hydrogel, in front of a substantially similar release rate. However, the particular pH-dependent behavior of alginate reticulated with Mn^{2+} can be potentially exploited for pH-triggered drug release. The idea, hereby suggested for the first time and qualitatively demonstrated, is that a device coated with Mn^{2+} reticulated alginate can almost instantaneously release all its drug load when it reaches an environment characterized by an acidic pH (e.g., the stomach). Indeed, acid degradable polymers present interesting properties for drug delivery [39,40]. Diffusion-driven discharge of a drug from a hydrogel is governed by n-order release kinetics [41], which are in the vast majority of the cases desirable to achieve sustained release. Degradable materials, however, are highly attractive when steep release gradients must be provided (with almost instantaneous release of the drug) [42]. This is the case for drugs that require high instantaneous concentrations to carry out their therapeutic action. Considering a possible application for pH-triggered drug release, the remaining experimentation presented in the paper was carried out on Mn-containing alginate devices.

3.5. Magnetic Actuation

Prior to functionality evaluation, the magnetic maneuverability of the devices was evaluated. Samples coated with Mn-reticulated alginate were placed inside a water-filled basin (to avoid hydrogel desiccation) and actuated, applying a rotating magnetic field. The principle exploited to move the microrobots is the continuous application of torque on the magnetic material present in the devices to generate a rolling motion. This actuation route, employed also in our previous works [22,33], requires the presence of a permanently magnetizable material. Consequently, an alloy like CoNiP (hard magnetic at the composition employed) was selected. As visible in Figure S9, which depicts the VSM performed on a single device, CoNiP presented a remanence in the order of 1124 Oe (along the 0° direction) and 936 Oe (along the 90° direction). In the convention employed for the VSM test, 0° refers to the direction parallel to the symmetry axis of the device, while 90° is the direction perpendicular to the same axis. These high remanence values are a consequence of the composition of the alloy: 3.08% wt. P, 84.75% wt. Co and 12.24% wt. Ni. Electrodeposited CoNiP layers presenting this composition are characterized by a typical hexagonal close-packed (hcp) structure with a marked preferential orientation along the (002), which justify the high level of coercivity and remanence achievable [43].

During rolling actuation, the permanent magnetization vector M introduced in the device by placing it in contact with a permanent NdFeB magnet continuously aligns itself (Figure 5a) with the external rotating field B. In this way, the device rotates around its symmetry axis (marked with a dashed black line in Figure 5a). The contact with a solid surface generates a net forward motion on the device. If the orientation of the axis with B rotates is varied, the direction of the motion can be adjusted and the device can be steered (Figure 5a). Obviously, in analogy with a wheel, there is a direct dependency between the rotation frequency of the magnetic field and the forward speed of the device. Under the assumption of an ideal contact between the device and the substrate, the speed of the first (v) correlates to the frequency (θ) according to Equation (9).

$$v = 2\pi r\,\theta \tag{9}$$

r corresponds to the radius of the device (comprising the rigid part of the device and the hydrogel layer deposited on top). Consequently, the speed/frequency relationship should be linear at any frequency. As demonstrated in our previous paper [32], this is never true for hydrogel-coated devices due to the presence of the hydrogel itself. Figure 5b reports the results obtained by performing a linear actuation on a device at increasing frequency. The data clearly evidenced that the uncoated device almost perfectly followed a linear behavior (evidenced by the fitting performed). The hydrogel-coated device, on the contrary, deviated from the linearity at frequencies higher than 2 Hz. The linear fitting was, in this case, limited to the 0 to 2 Hz range.

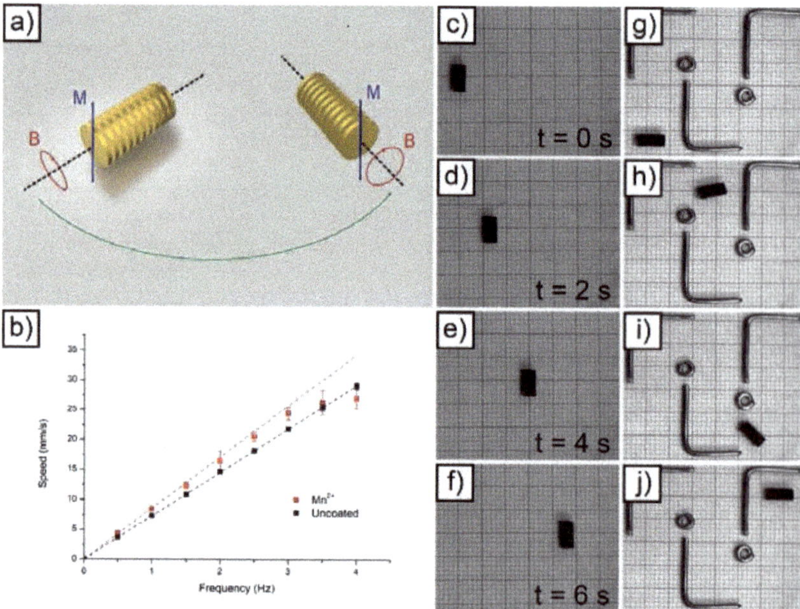

Figure 5. Schematic representation of the magnetic actuation principle employed for microrobots actuation (**a**); speed vs. frequency relationship for uncoated and Mn alginate-coated microdevices (**b**); linear actuation of a single Mn alginate-coated device (from (**c–f**)); remote guidance of a Mn alginate-coated device to avoid metallic obstacles (from (**g–j**)).

Considering Equation (9), it appears evident that the slope of the fitted curves visible in Figure 5b corresponds to $2\pi r$. Consequently, the radius of the devices (resulting from the sum between the radius of rigid platform and the radius of the hydrogel layer applied on top) was easily extrapolated from the slope of the fitted curves. The result was equal to 1.161 mm for the uncoated device and to 1.364 mm for the coated device. The value obtained for the uncoated microrobot is in line with the theoretical one reported in Table S1 (1.1975 mm). The value obtained in the case of the coated device suggests that the thickness of the Mn-reticulated hydrogel layer was equal to 203 µm. This value does not correspond to the absolute value of thickness for the hydrogel coating. Inside the plates, for example, this value was realistically different due to hydrogel accumulation. The value obtained represents a mean hydrogel thickness in correspondence with the external radius of the devices (on the parallel plates).

Figure 5c–f and Supporting Video S1 depict the linear actuation performed to obtain the speed data at 0.5 Hz in Figure 5b. The linear correlation between elapsed time and covered length was evidenced by showing four different frames acquired at increasing times. Figure 5g–j and Supporting Video S2 report the result obtained by guiding a Mn-reticulated alginate-coated device to avoid metallic obstacles placed inside the water-filled basin. The device evidenced an excellent maneuverability at 0.5 Hz actuation frequency.

3.6. Targeted pH Dependent Release

To experimentally validate the basic concept of targeted pH-dependent delivery from a qualitative point of view, two distinct tests were carried out. Initially, the capacity of targeting drug delivery of the devices was tested. A Mn-reticulated alginate-coated device, suitably loaded with RhB, was placed inside a basin and actuated. UV light was exploited to evidence, thanks to RhB fluorescence at 345 nm, the presence of the drug on the walls of the basin. The result obtained is presented in Figure 6a–f and in Supporting Video S3. At the beginning of the test, the device moved towards the center of the basin along a curved trajectory. Once it reached the center of the basin, the device was left there for a few seconds to allow release. Then, the device was moved, always along a curved trajectory, toward the lower corner of the basin. Besides the tracks left in the water, UV light evidenced the presence of RhB adsorbed on the bottom of the basin in correspondence with the point where the device stayed for a few seconds. This qualitative result suggests the idea that these devices could be potentially guided in vivo inside the gastrointestinal tract, left in contact with the inner parts of the organs to locally favor the absorption of a drug of choice.

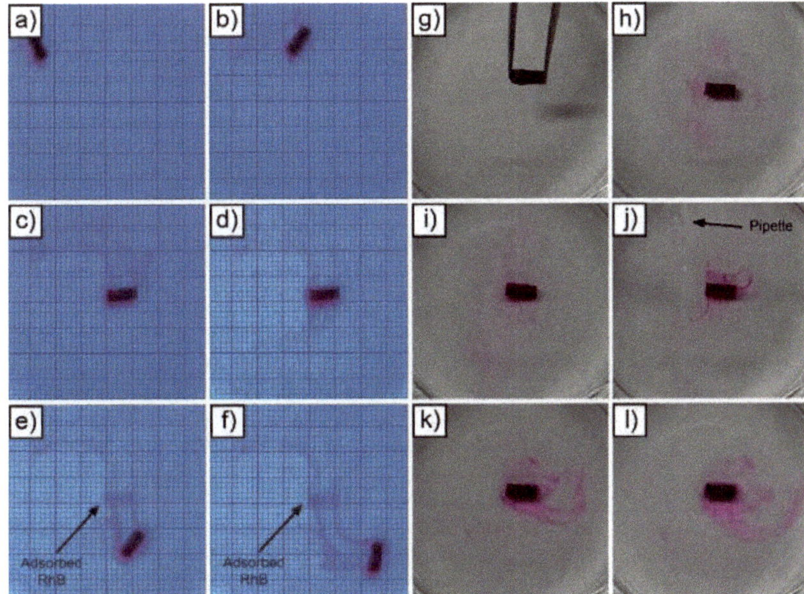

Figure 6. Targeted drug delivery from a device (from (**a–f**)); Mn-reticulated hydrogel dissolution test at pH 3 (from (**g–l**)).

In the second test, visualized in Figure 6g–l and in Supporting Video S4, a device analogous to the one used for the previous test was immersed in water. As expected, the microrobot immediately started releasing RhB according to a Fickian diffusion mechanism. In these conditions, nearly total release of the drug potentially required almost two hours (as visible in Figure 4a). To exponentially increase the release rate, thus allowing immediate drug release, the pH of the water-filled basin was artificially lowered by adding HCl. Specifically, the stomach environment was simulated, reaching pH 3 [44]. To obtain this condition, a suitable quantity of HCl was added to the basin and rapidly mixed with a pipette. Under these conditions, the device started to quickly release all the RhB contained inside the alginate layer, which rapidly disintegrated in a few-minute timeframe.

Thanks to the low pH, the device expelled its RhB load almost immediately, with a drug release not governed by diffusion in the hydrogel. Furthermore, hydrogel removal by HCl allowed direct reuse of the device.

3.7. Possible Toxicity of Mn Reticulated Alginate

Possible dose-dependent Mn toxicity is a matter of major importance when evaluating the potential use of Mn-reticulated alginate for drug delivery applications. In fact, when the acidity mechanically degrades the alginate layer, all the Mn contained is immediately released in the environment. Mn is an element required in small quantities by the human body [45], but its excess may potentially result in a condition known as manganism. Mn is normally obtained through dietary intake and is readily absorbed by the gastrointestinal apparatus. O'Neal et al. evidenced that the average daily intake for many Western diets is between 2.3 and 8.8 mg [45]. Conversely, the WHO standard for drinking water is 400 μg/L of Mn [45]. These values, which represent safe Mn consumption from food and drinking water, can be compared with the mean amount of Mn released by a device when the alginate layers dissolve. Each device is coated with 14 ± 0.5 mg of Mn-reticulated hydrogel (data from Figure 3). According to the supplier (Sigma Aldrich), the mean molecular weight of the alginate employed ($MW_{hydrogel}$) was between 120 and 190 kDa (120 kDa was considered). The total amount of Mn within every device can be obtained considering the number of monomeric units present in 14 mg of the hydrogel layer. The percentage of polymer with respect to water in these hydrogels is approximately 20% wt., so the number of monomers is approximately 8×10^{18}. Taking into consideration the ratio between one atom of Mn every two monomeric units of alginate, we can obtain the Mn amount, which is around 350 μg. This value is lower than the tolerable amount recommended by the WHO for a liter of drinking water (400 μg). In front of these realistic considerations, the modest amount of Mn released from a single device cannot be considered harmful for the human body.

4. Discussion

The experiments carried out in the present work expanded the possibilities for targeted drug release from hydrogel layers loaded on magnetically moveable rigid platforms. Initially, the hydrogel loading capability of the devices was optimized by introducing a honey dipper design. This approach resulted in a 20% load increase with respect to our previous design. Microdevices were successfully 3D printed via stereolithography and wet metallized with a sequence of functional layers. The use of barrel plating allowed to deposit thick metallic layers of CoNiP and Au without direct clamping of the single device (which would have resulted in uncoated areas). After plating, the devices were successfully coated with a controllable amount of alginate hydrogel. The type of crosslinker employed for reticulation was found to have a remarkable influence on the amount of hydrogel deposited on a single device, with Mn allowing the lowest amount of loaded hydrogel. On the contrary, the three different reticulating agents evidenced a limited effect on drug release performances. Mn alginate-coated devices were actuated applying external rotating magnetic fields, evidencing a good control over speed and position. Even though less advantageous in terms of loadable hydrogel amount, Mn-reticulated alginate was characterized by an interesting property: the layer dissolved at low pH, immediately releasing all its drug load. The possible applicability of this property, which can potentially be exploited to trigger an instantaneous release by mean of pH variation, was successfully demonstrated. A single device bearing Mn-reticulated hydrogel was subjected to a controlled pH variation simulating the gastric environment. The result obtained qualitatively evidenced the possibility of pH-triggering an instantaneous release, characterized by steep release rates and high instantaneous drug bioavailability. These features may be useful for specific types of drugs that present high active concentrations. The results obtained suggest the idea that hydrogels reticulated with ions alternative to the classically employed Ca can be used to implement a pH-dependent release. They constitute a significant base for further experiments, which are currently under evaluation. Specifically, Mn-reticulated layers could be coupled with Ca-reticulated ones to tune the release. In addition, alternative small-sized biocompatible ions (or mixtures of different ions) could be evaluated to tune the pH sensitivity of the hydrogel layer.

Supplementary Materials: The following are available online at https://www.mdpi.com/article/10.3390/technologies9020043/s1, Table S1: Theoretical and experimental dimensions of the microdevices, Figure S1: Honey dipper, Figure S2: dimensions of the devices, Figure S3: SEM image of an as printed device (10,000 X), Figure S4: Section of a metallized device (optical microscope), Figure S5: SEM image of a metallized device (10,000 X), Figure S6: EDS analysis of the surface depicted in Figure S5, Figure S7: Hydrogel loading procedure, Figure S8: Sample holder, Figure S9: VSM of an uncoated device, Supplementary Video S1, Supplementary Video S2, Supplementary Video S3, Supplementary Video S4.

Author Contributions: Conceptualization: R.B. and F.R.; methodology: R.B. and F.R.; validation: R.B., F.R., A.R. and F.P.; investigation: R.B., F.P., A.R., R.P. and A.N.; resources: M.L.; writing—original draft preparation: R.B. and F.R.; writing—review and editing: R.B., F.R. and A.N.; supervision: R.B. and F.R. All authors have read and agreed to the published version of the manuscript.

Funding: This research received no external funding.

Institutional Review Board Statement: Not applicable.

Informed Consent Statement: Not applicable.

Data Availability Statement: The data that support the findings of this study are available from the corresponding author, R. Bernasconi, upon reasonable request.

Acknowledgments: The authors wish to acknowledge the support given by Salvador Panè (ETH Zurich), who provided access to the Octomag magnetic manipulation system and to the VSM setup employed in the paper.

Conflicts of Interest: The authors declare no conflict of interest.

References

1. Brater, D.C.; Daly, W.J. Clinical pharmacology in the Middle Ages: Principles that presage the 21st century. *Clin. Pharmacol. Ther.* **2000**, *67*, 447–450. [CrossRef]
2. Abu-Thabit, N.Y.; Makhlouf, A.S.H. *Historical Development of Drug Delivery Systems: From Conventional Macroscale to Controlled, Targeted, and Responsive Nanoscale Systems*; Elsevier: Amsterdam, The Netherlands, 2018; ISBN 9780081019979.
3. Wang, B.; Hu, L.; Siahaan, T.J. *Drug Delivery: Principles and Applications*; John Wiley & Sons: Hoboken, NJ, USA, 2016; ISBN 1118833236.
4. Sekyere, J.O.; Asante, J. Emerging mechanisms of antimicrobial resistance in bacteria and fungi: Advances in the era of genomics. *Future Microbiol.* **2018**, *13*, 241–262. [CrossRef]
5. Vasan, N.; Baselga, J.; Hyman, D.M. A view on drug resistance in cancer. *Nature* **2019**, *575*, 299–309. [CrossRef] [PubMed]
6. Holowka, E.P.; Bhatia, S.K. Smart Drug Delivery Systems. In *Drug Delivery*; Springer: Berlin/Heidelberg, Germany, 2014; pp. 265–316.
7. Alvarez-Lorenzo, C.; Concheiro, A. Smart drug delivery systems: From fundamentals to the clinic. *Chem. Commun.* **2014**, *50*, 7743–7765. [CrossRef] [PubMed]
8. Lee, Y.; Thompson, D.H. Stimuli-responsive liposomes for drug delivery. *Wiley Interdiscip. Rev. Nanomed. Nanobiotechnol.* **2017**, *9*, e1450. [CrossRef] [PubMed]
9. Pattni, B.S.; Chupin, V.V.; Torchilin, V.P. New developments in liposomal drug delivery. *Chem. Rev.* **2015**, *115*, 10938–10966. [CrossRef] [PubMed]

10. Zelikin, A.N. Drug releasing polymer thin films: New era of surface-mediated drug delivery. *ACS Nano* **2010**, *4*, 2494–2509. [CrossRef] [PubMed]
11. Hu, Q.; Li, H.; Wang, L.; Gu, H.; Fan, C. DNA nanotechnology-enabled drug delivery systems. *Chem. Rev.* **2018**, *119*, 6459–6506. [CrossRef] [PubMed]
12. Chauhan, A.S. Dendrimers for drug delivery. *Molecules* **2018**, *23*, 938. [CrossRef]
13. Zhou, Q.; Zhang, L.; Yang, T.; Wu, H. Stimuli-responsive polymeric micelles for drug delivery and cancer therapy. *Int. J. Nanomed.* **2018**, *13*, 2921. [CrossRef]
14. Manzano, M.; Vallet-Regí, M. Mesoporous silica nanoparticles for drug delivery. *Adv. Funct. Mater.* **2020**, *30*, 1902634. [CrossRef]
15. Vangijzegem, T.; Stanicki, D.; Laurent, S. Magnetic iron oxide nanoparticles for drug delivery: Applications and characteristics. *Expert Opin. Drug Deliv.* **2019**, *16*, 69–78. [CrossRef]
16. Ghitman, J.; Biru, E.I.; Stan, R.; Iovu, H. Review of hybrid PLGA nanoparticles: Future of smart drug delivery and theranostics medicine. *Mater. Des.* **2020**, *193*, 108805. [CrossRef]
17. Nelson, B.J.; Kaliakatsos, I.K.; Abbott, J.J. Microrobots for Minimally Invasive Medicine. *Annu. Rev. Biomed. Eng.* **2010**, *12*, 55–85. [CrossRef]
18. Chen, X.-Z.; Hoop, M.; Mushtaq, F.; Siringil, E.; Hu, C.; Nelson, B.J.; Pané, S. Recent developments in magnetically driven micro-and nanorobots. *Appl. Mater. Today* **2017**, *9*, 37–48. [CrossRef]
19. Jang, D.; Jeong, J.; Song, H.; Chung, S.K. Targeted drug delivery technology using untethered microrobots: A review. *J. Micromech. Microeng.* **2019**, *29*, 53002. [CrossRef]
20. Erkoc, P.; Yasa, I.C.; Ceylan, H.; Yasa, O.; Alapan, Y.; Sitti, M. Mobile Microrobots for Active Therapeutic Delivery. *Adv. Ther.* **2019**, *2*, 1800064. [CrossRef]
21. Xu, T.; Yu, J.; Yan, X.; Choi, H.; Zhang, L. Magnetic actuation based motion control for microrobots: An overview. *Micromachines* **2015**, *6*, 1346–1364. [CrossRef]
22. Bernasconi, R.; Cuneo, F.; Carrara, E.; Chatzipirpiridis, G.; Hoop, M.; Chen, X.; Nelson, B.J.; Pané, S.; Credi, C.; Levi, M.; et al. Hard-magnetic cell microscaffolds from electroless coated 3D printed architectures. *Mater. Horiz.* **2018**, *5*, 699–707. [CrossRef]
23. Bernasconi, R.; Favara, N.; Fouladvari, N.; Invernizzi, M.; Levi, M.; Vidal, S.P.; Magagnin, L. Nanostructured Polypyrrole Layers Implementation on Magnetically Navigable 3D Printed Microdevices for Targeted Gastrointestinal Drug Delivery. *Multifunct. Mater.* **2020**, *3*, 045003. [CrossRef]
24. Jeon, S.; Kim, S.; Ha, S.; Lee, S.; Kim, E.; Kim, S.Y.; Park, S.H.; Jeon, J.H.; Kim, S.W.; Moon, C. Magnetically actuated microrobots as a platform for stem cell transplantation. *Sci. Robot.* **2019**, *4*, eaav4317. [CrossRef]
25. Cabanach, P.; Pena-Francesch, A.; Sheehan, D.; Bozuyuk, U.; Yasa, O.; Borros, S.; Sitti, M. Zwitterionic 3D-Printed Non-Immunogenic Stealth Microrobots. *Adv. Mater.* **2020**, *32*, 2003013. [CrossRef]
26. Li, J.; Pumera, M. 3D printing of functional microrobots. *Chem. Soc. Rev.* **2021**, *50*, 2794–2838. [CrossRef]
27. Lee, H.; Kim, D.; Kwon, S.; Park, S. Magnetically Actuated Drug Delivery Helical Microrobot with Magnetic Nanoparticle Retrieval Ability. *ACS Appl. Mater. Interfaces* **2021**, *13*, 19633–19647. [CrossRef]
28. Park, J.; Kim, J.; Pané, S.; Nelson, B.J.; Choi, H. Acoustically Mediated Controlled Drug Release and Targeted Therapy with Degradable 3D Porous Magnetic Microrobots. *Adv. Healthc. Mater.* **2020**, *10*, 2001096. [CrossRef]
29. Lee, S.; Kim, J.; Kim, J.; Hoshiar, A.K.; Park, J.; Lee, S.; Kim, J.; Pané, S.; Nelson, B.J.; Choi, H. A Needle-Type Microrobot for Targeted Drug Delivery by Affixing to a Microtissue. *Adv. Healthc. Mater.* **2020**, *9*, 1901697. [CrossRef] [PubMed]
30. Narayanaswamy, R.; Torchilin, V.P. Hydrogels and their applications in targeted drug delivery. *Molecules* **2019**, *24*, 603. [CrossRef] [PubMed]
31. Sood, N.; Bhardwaj, A.; Mehta, S.; Mehta, A. Stimuli-responsive hydrogels in drug delivery and tissue engineering. *Drug Deliv.* **2016**, *23*, 748–770. [CrossRef] [PubMed]
32. Bernasconi, R.; Mauri, E.; Rossetti, A.; Rimondo, S.; Suriano, R.; Levi, M.; Sacchetti, A.; Pané, S.; Magagnin, L.; Rossi, F. 3D integration of pH-cleavable drug-hydrogel conjugates on magnetically driven smart microtransporters. *Mater. Des.* **2021**, *197*, 109212. [CrossRef]
33. Bernasconi, R.; Carrara, E.; Hoop, M.; Mushtaq, F.; Chen, X.; Nelson, B.J.; Pané, S.; Credi, C.; Levi, M.; Magagnin, L. Magnetically navigable 3D printed multifunctional microdevices for environmental applications. *Addit. Manuf.* **2019**, *28*, 127–135. [CrossRef]
34. Da Silva Fernandes, R.; Tanaka, F.N.; de Moura, M.R.; Aouada, F.A. Development of alginate/starch-based hydrogels crosslinked with different ions: Hydrophilic, kinetic and spectroscopic properties. *Mater. Today Commun.* **2019**, *21*, 100636. [CrossRef]
35. Rizwan, M.; Yahya, R.; Hassan, A.; Yar, M.; Azzahari, A.D.; Selvanathan, V.; Sonsudin, F.; Abouloula, C.N. pH sensitive hydrogels in drug delivery: Brief history, properties, swelling, and release mechanism, material selection and applications. *Polymers* **2017**, *9*, 137. [CrossRef] [PubMed]
36. Kummer, M.P.; Abbott, J.J.; Kratochvil, B.E.; Borer, R.; Sengul, A.; Nelson, B.J. Octomag: An electromagnetic system for 5-DOF wireless micromanipulation. *IEEE Trans. Robot.* **2010**, *26*, 1006–1017. [CrossRef]
37. Dodero, A.; Pianella, L.; Vicini, S.; Alloisio, M.; Ottonelli, M.; Castellano, M. Alginate-based hydrogels prepared via ionic gelation: An experimental design approach to predict the crosslinking degree. *Eur. Polym. J.* **2019**, *118*, 586–594. [CrossRef]
38. Blandino, A.; Macias, M.; Cantero, D. Formation of calcium alginate gel capsules: Influence of sodium alginate and $CaCl_2$ concentration on gelation kinetics. *J. Biosci. Bioeng.* **1999**, *88*, 686–689. [CrossRef]

39. Binauld, S.; Stenzel, M.H. Acid-degradable polymers for drug delivery: A decade of innovation. *Chem. Commun.* **2013**, *49*, 2082–2102. [CrossRef] [PubMed]
40. Uhrich, K.E.; Abdelhamid, D. *Biodegradable and Bioerodible Polymers for Medical Applications*; Elsevier: Amsterdam, The Netherlands, 2016; ISBN 9781782421139.
41. Bruschi, M.L. *Mathematical Models of Drug Release*; Woodhead Publishing: Sawston, UK, 2015; ISBN 9780081000922.
42. Lao, L.L.; Peppas, N.A.; Boey, F.Y.C.; Venkatraman, S.S. Modeling of drug release from bulk-degrading polymers. *Int. J. Pharm.* **2011**, *418*, 28–41. [CrossRef]
43. Park, D.Y.; Myung, N.V.; Schwartz, M.; Nobe, K. Nanostructured magnetic CoNiP electrodeposits: Structure-property relationships. *Electrochim. Acta* **2002**, *47*, 2893–2900. [CrossRef]
44. Russell, T.L.; Berardi, R.R.; Barnett, J.L.; Dermentzoglou, L.C.; Jarvenpaa, K.M.; Schmaltz, S.P.; Dressman, J.B. Upper Gastrointestinal pH in Seventy-Nine Healthy, Elderly, North American Men and Women. *Pharm. Res.* **1993**, *10*, 187–196. [CrossRef]
45. O'Neal, S.L.; Zheng, W. Manganese Toxicity Upon Overexposure: A Decade in Review. *Curr. Environ. Health Rep.* **2015**, *2*, 315–328. [CrossRef]

 *technologies*

Review

Three-Dimensional Printed Models for Preoperative Planning and Surgical Treatment of Chest Wall Disease: A Systematic Review

Beatrice Leonardi [1], Annalisa Carlucci [1], Antonio Noro [1], Mary Bove [1], Giovanni Natale [1], Giorgia Opromolla [1], Rosa Mirra [1], Davide Pica [1], Francesca Capasso [1], Vincenzo Di Filippo [1], Gaetana Messina [1], Francesco Ferrigno [2], Anna Cecilia Izzo [1], Giovanni Vicidomini [1], Mario Santini [1] and Alfonso Fiorelli [1,*]

[1] Thoracic Surgery Unit, University of Campania Luigi Vanvitelli, FCCP Via Pansini, 83100 Naples, Italy; beatriceleonardi01@gmail.com (B.L.); annalisacarlucci88@gmail.com (A.C.); ant.noro@gmail.com (A.N.); bovemary10@gmail.com (M.B.); dott.natale.giovanni@gmail.com (G.N.); giorgia.opromolla@studenti.unicampania.it (G.O.); rosamirra92@yahoo.it (R.M.); davidepica@gmail.com (D.P.); francesca.capasso93@gmail.com (F.C.); vincenzodifilippo16@libero.it (V.D.F.); adamessina@virgilio.it (G.M.); annaizzo@libero.it (A.C.I.); giovanni.vicidomini@unicampania.it (G.V.); mario.santini@unicampania.it (M.S.)
[2] Pneumology Unit, Mauro Scarlato Hospital, 84018 Scafati, Italy; francescoferrigno@libero.it
*   Correspondence: alfonso.fiorelli@unicampania.it; Tel.: +39-3381030061; Fax: +39-0815665230

Citation: Leonardi, B.; Carlucci, A.; Noro, A.; Bove, M.; Natale, G.; Opromolla, G.; Mirra, R.; Pica, D.; Capasso, F.; Di Filippo, V.; et al. Three-Dimensional Printed Models for Preoperative Planning and Surgical Treatment of Chest Wall Disease: A Systematic Review. *Technologies* **2021**, *9*, 97. https://doi.org/10.3390/technologies9040097

Academic Editor: Roberto Bernasconi

Received: 22 October 2021
Accepted: 19 November 2021
Published: 3 December 2021

Publisher's Note: MDPI stays neutral with regard to jurisdictional claims in published maps and institutional affiliations.

Abstract: Introduction: In chest wall reconstruction, the main objectives are the restoration of the chest wall integrity, function, and aesthetic, which is often achieved with the placement of implants. We aimed to evaluate whether 3D printed models can be useful for preoperative planning and surgical treatment in chest wall reconstruction to improve the outcome of the surgery and to reduce the rate of complications. Methods: We conducted a systematic review of literature using PubMed, Scopus, Embase, and Google Scholar databases until 8 November 2021 with the following keywords: ("3D printing" or "rapid prototyping" or "three-dimensional printing" or "bioprinting") and ("chest wall" or "rib" or "sternum" or "ribcage" or "pectus excavatum"). Results were then manually screened by two independent authors to select studies relevant to 3D printing application in chest wall reconstruction. The primary outcome was morphological correction, and secondary outcomes were changes in operating time and procedure-related complication rate. Results: Eight articles were included in our review. Four studies were related to pectus excavatum correction, two studies were related to rib fracture stabilization, and two studies were related to chest wall tumor resection and reconstruction. Seven studies reported 3D printing of a thorax model or template implants for preoperative planning and implant modeling, and one study reported 3D printing of a PEEK prosthesis for direct implantation. Four studies reported comparison with a conventionally treated control group, and three of them detected a shorter operative time in the 3D printing model-assisted group. Satisfactory morphological correction was reported in all studies, and six studies reported a good implant fitting with minimal need for intraoperative adjustments. There were no major intraoperative or postoperative complications in any of the studies. Conclusions: The use of 3D printing models in chest wall reconstruction seems to be helpful for the production of personalized implants, reducing intraoperative adjustments. Results of morphological correction and postoperative recovery after the 3D printing-assisted surgery were satisfactory in all studies with a low rate of complication. Our literature review suggests good results regarding prosthesis fitting, accuracy of surgical planning, and reduction in operative time in 3D printing-assisted procedures, although more evidence is needed to prove this observation.

Keywords: 3D printing; chest wall; surgery

1. Introduction

Chest wall reconstruction is required in different situations, such as traumatic injuries, thoracic tumors, or chest wall malformations. The main objectives in reconstructing the thoracic wall are restoring the chest wall integrity and rigidity, preserving the physiological function of the thorax, and protecting internal organs, while pursuing a good aesthetic result [1].

Conventional methods for chest wall reconstruction are represented by prosthetic replacements using biologic (allografts and homografts), synthetic (methyl methacrylate, PTFE, polypropylene), or metallic (titanium, stainless steel) meshes, bars, or plates, depending on the pathology of interest [2].

The prosthesis fitting is crucial to achieve all of the mentioned objectives, and 3D reconstruction based on CT images offers the possibility to accurately study the pathological anatomy of the chest wall, thus allowing the procedure to be meticulously planned and the prosthetic materials to be provided in advance [3]. The application of 3D reconstruction is implemented in preexisting imaging techniques that are commonly used in chest wall visualization. These are mostly CT, but also MRI and US, when appropriate in preoperative study or postoperative evaluation of complications.

In the surgical field, 3D printing has been embraced as a tool for preoperative planning, surgical simulation and training [4], patient education [5], and to produce templates or molds for surgical implants [6,7] for 3D printing prosthesis and implants (especially in orthopedic and maxillofacial surgery) [8] and in tissue engineering (3D bioprinting) [9].

Due to its recent development in the medical field, reports regarding the use of 3D printing in thoracic surgery, and specifically in chest wall reconstruction, are still limited. Nonetheless, many aspects of 3D printing-assisted procedures are of interest, including the possibility to better understand the anatomy of complex situations, such as large thoracic tumors, which often invade adjacent structures, thus making ita challenge for the surgeon to perform a successful resection while needing to achieve R0 resection margins. Another interesting application is the possibility to project custom-made implants, specifically built on the anatomy of the patient, and to rehearse the procedure on a customized model.

Analyzing current evidence on this matter, we aimed to evaluate whether 3D printed models could be useful for preoperative planning and surgical treatment in chest wall reconstruction to improve the outcome of the surgery and to reduce the rate of complications.

2. Materials and Methods

2.1. Search Strategy

We conducted a systematic review according to the PRISMA protocol [10] to assess the current use of 3D printing in chest wall reconstruction. Four medical databases (PubMed, Scopus, Embase, and Google Scholar, updated to 8 November 2021) were searched to collect studies regarding the subject of interest using the following keywords: ("3D printing" or "rapid prototyping" or "three-dimensional printing" or "bioprinting") and ("chest wall" or "rib" or "sternum" or "ribcage" or "pectus excavatum"). We conducted an additional manual search of the bibliographies of the selected articles. The details of the database search are reported in Table 1.

2.2. Selection Process

We included in our review: (I) papers published in English; (II) studies that reported surgical application of 3D printing models for preoperative planning or surgical reconstruction of the chest wall; (III) original research articles. The exclusion criteria were as follows: (I) papers not published in English; (II) case reports, case series including fewer than 5 patients, reviews, abstracts, meta-analyses, brief communications, editorials, and letters; (III) papers based on the same study population, in which case we selected the most recent article to avoid duplication.

Table 1. Details of the database search.

Database	Search Results (8 November 2021)	Keywords
Pubmed	143	(3D printing OR rapid prototyping OR three-dimensional printing OR bioprinting) AND (chest wall OR rib OR sternum OR ribcage OR pectus excavatum)
Scopus	244	("3D printing" OR "rapid prototyping" OR "three-dimensional printing" OR "bioprinting") AND ("chest wall" OR "rib" OR "sternum" OR "ribcage" OR "pectus excavatum")
Embase	182	(3D printing OR rapid prototyping OR three-dimensional printing OR bioprinting) AND (chest wall OR rib OR sternum OR ribcage OR pectus excavatum)
Google Scholar	139	(3D printing or three-dimensional printing) AND (chest wall or rib or sternum or pectus excavatum)

Two independent reviewers (BL, AN) analyzed the search results, and proceeded to inspect the articles' titles and abstracts to exclude papers that were not related to the topic of interest. Full text of the remaining articles was then evaluated to determine if they met our inclusion/exclusion criteria. Disagreements were resolved through discussion with two senior reviewers (AF, MS).

The primary outcome of our review was morphological correction, secondary outcomes were changes in operating time and procedure-related complication rate.

2.3. Quality and Risk of Bias Assessment

The quality and risk of bias for each study was independently evaluated using the Downs and Black assessment checklist [11] by two authors (BL, AN). Differences between the two reviewers were solved through discussion with a third author (AF). The total score for this 27-item checklist ranges from 0 to 28 points. We considered scores as follows: Excellent (24–28 points), Good (19–23 points), Fair (14–18), Poor (<14 points). The synthetized result of the assessment is reported in Table 2, indicating an overall good methodological quality of the considered studies.

Table 2. Quality assessment results (Downs and Black checklist).

Study	Reporting (Max 11)	External Validity (Max 3)	Internal Validity/Bias (Max 7)	Internal Validity/Confounding (Max 6)	Power (Max 1)	Total (Max 28)
Zhou et al.	10	3	5	3	0	21
Bellia Munzon et al.	10	3	5	3	1	22
Wang et al. (2020)	6	1	5	3	0	15
Wu et al.	11	3	5	3	0	22
Wang et al. (2019)	9	3	5	3	0	20
Huang et al.	11	3	5	3	1	23
Chen et al.	11	3	5	3	1	23
Gaspar Pérez et al.	10	3	5	3	0	21

3. Results

A total of 708 titles were identified via the search of the previously mentioned databases, whereas no additional study of interest was found via a manual search of the bibliographies of the selected studies (Figure 1). We subsequently excluded 299 papers as duplicates.

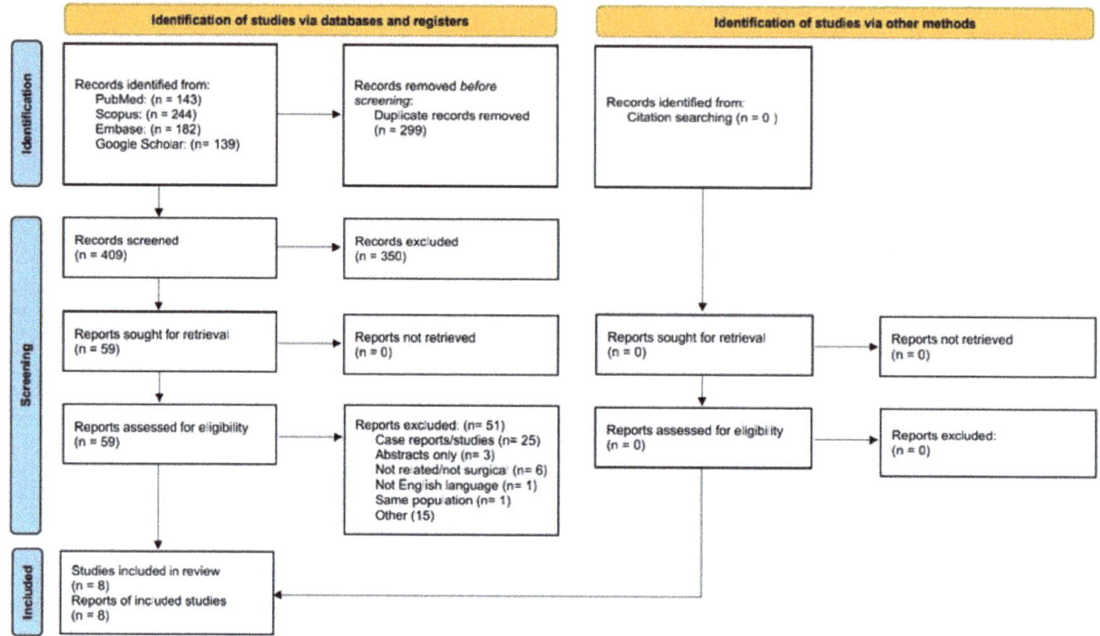

Figure 1. Flow chart of the study according to PRISMA guidelines.

Based on the titles and abstracts, the remaining 409 papers were evaluated, and 350 of these were excluded as the object was not related to the topic of our research. The remaining 59 articles were analyzed through full-text examination by all authors, who further excluded 51 studies. Finally, a total of eight papers were selected, which constitute the subject of our systematic review.

The following data were extracted from the selected papers, as summarized in Table 3: the authors, the year of publication, the country, the study design, the number of patients that received a 3D printing assisted procedure, the application, the disease of interest, the study limitations, the structure printed, the processing software, the 3D printer, the materials.

Table 3. Characteristics of the studies evaluated.

Authors, Year of Publication, Country	Study Design	N° of Patients	Application	Disease of Interest	Structure Printed	CT-Images Processing Software	3D Printer	Materials	Limitations
Zhou et al. (7), 2021, China	Prospective	16	Surgical planning and prosthetic modeling	Rib fractures	Ribcage model	MDT2AB-010A, Meditool Medical Technology (Shanghai)	pangu4.1, Meditool Medical Technology	Photosensitive resin	No conventionally treated control group Small sample size
Bellia-Munzon et al. (6) 2020, Argentina	Prospective	130	Surgical planning and prosthetic modeling	Pectus excavatum	Customized implant bars template	Erkom 3D Chest Wall Pro 1.0, Pampamed (Buenos Aires)	ERKOM 3D	Polyvinyl acetate	Non controlled study nature

Table 3. *Cont.*

Authors, Year of Publication, Country	Study Design	N° of Patients	Application	Disease of Interest	Structure Printed	CT-Images Processing Software	3D Printer	Materials	Limitations
Zhou et al. (7), 2021, China	Prospective	16	Surgical planning and prosthetic modeling	Rib fractures	Ribcage model	MDT2AB-010A, Meditool Medical Medical Technology (Shanghai)	pangu4.1, Meditool Medical Technology	Photosensitive resin	No conventionally treated control group Small sample size
Bellia-Munzon et al. (6) 2020, Argentina	Prospective	130	Surgical planning and prosthetic modeling	Pectus excavatum	Customized implant bars template	Erkom 3D Chest Wall Pro 1.0, Pampamed (Buenos Aires)	ERKOM 3D	Polyvinyl acetate	Non controlled study nature
Wang et al. (9) 2020, China	Prospective	6	Surgical planning and prosthetic modeling	Pectus excavatum	Ribcage model	3D-DOCTOR, Able Software Corp (Lexington)	Not specified	Polylactic acid	No conventionally treated control group Small sample size
Wu et al. (11), 2018, China	Retrospective	6	Surgical planning and prosthetic modeling	Chest wall tumor	Chest wall tumor model	Amira Thermo Fisher Scientific (Berlin)	Formlabs Form2 /MakerBot Replica-torTM 2X	Liquid photosensitive resin	Retrospective nature Small sample size
Wang et al. (11), 2019, China	Prospective	18	Prosthetic replacement	Chest wall tumor	Ribs and sternum prostheses	Mimics 17.0, Materialise MV (Leuven)/ Geomagic Studio version 2012 3D Systems (Morrisville)	Jugao-AM-Doctor, Shaanxi Jugao-AM Technology	Polyetheretherketone	No conventionally treated control group Small sample size
Huang et al. (8), 2019, China	Retrospective	15	Surgical planning and prosthetic modeling	Pectus excavatum	Customized implant bars template	Meshmixer, Autodesk (San Rafael)	UP BOX, Beijing Tiertime Technology	Polylactic acid	Retrospective nature Small 3D printing group size
Chen et al. (5), 2018, China	Retrospective	16	Surgical planning and prosthetic modeling	Rib fractures	Ribcage model	Not specified	UP-BOX 3D printer, Denford	Acrylonitrile butadiene styrene	Retrospective nature Small sample size
Gaspar Pérez et al. (12), 2021, Spain	Prospective	6	Surgical planning and prosthetic modeling	Pectus excavatum	Ribcage model, customized implant bars template	Mimics 21.0/3-matic 13.0, Materialise MV (Leuven)	Ultimaker S5 3D, Ultimaker B.V.	Polylactic acid	No conventionally treated control group Small sample size

3.1. Study and Patient Characteristics

The eight studies included in our review were published between the year 2018 and 2021; six studies were from China, one was from Spain, and one was from Argentina. There were five prospective studies and three retrospective studies. Four studies were related to pectus excavatum correction, two to rib fracture stabilization, and two studies to chest wall tumor resection and reconstruction.

In all studies, 3D printing models were used for preoperative planning, whereas in only one study (Wang et al., 2019) it was also used for chest wall repair.

3.2. Rib Fracture Fixation

Zhou et al. [12] treated 16 patients with high complex rib fractures (identified as patients with one to three fractures in the 2nd to 4th ribs, with three fracture segments in each fractured site, the middle fracture segment $\leq$ 5 cm and presence of costal cartilage fracture) with framework locking-plate internal fixation combined with 3D printing. A 3D ribcage model based on 64-row spiral CT images was processed then imported into the 3D printer to obtain a true-size photosensitive resin model of the fractured ribs. The model was used to study the morphology of the fracture site, to select an appropriate incision site, and to simulate the fracture reduction. Placement, direction, length, and number of screws needed for fixation were planned and recorded based on the 3D printed rib model. During the surgery, there was no need for intraoperative adjustments to the locking plate. Follow-up at 5 to 10 months evaluated the recovery of the patients. A satisfactory chest wall reconstruction was achieved in all cases, confirmed by chest RX or CT scans after surgery.

Chen et al. [13] retrospectively evaluated the data of 48 patients who underwent surgical stabilization of rib fractures (SSRF). Patients were split into two groups based on whether or not 3D printing was used (32 patients in the conventional SSRF group vs. 16 in the 3D assisted SSRF group). Before surgery, in the 3D printing-assisted group, a 3D reconstruction was created from CT images (1 mm slice thickness), then a 3D printed model of the ribcage was produced in acrylonitrile butadiene styrene. Based on the model, the titanium plates were bent and cut to fit the patient's ribcage profile, and the length and number of the screws was recorded. The incision site was planned according to the 3D printed model. In comparison with the conventional SSRF group, the mean operative time (175.24 vs. 125 min $p = 0.003$,) and the mean operative time per fixed plate (52.99 vs. 35.41 min, $p < 0.001$) were significantly shorter in patients in the 3D printing SSRF group. The incision length (14.19 vs. 8.71 cm, $p = 0.002$) and the wound length per fixed rib (4.19 vs. 2.54 cm, $p < 0.001$) were significantly smaller in the 3D printing SSRF group. There was no statistically significant difference in hospital stay, ICU stay, and postoperative complications (empyema, broken plates, and wound infection happened in four patients in the SSRF group) between the two groups.

3.3. Pectus Excavatum Repair

Bellia-Munzon et al. [14] prospectively collected data of 130 patients who underwent minimally invasive repair of pectus excavatum (MIRPE) with the use of 3D printing. All patients underwent a CT scan with 3D reconstruction (1 mm slice thickness). A flexed 3D printed template of the implant bars was elaborated based on the CT images and clinical information of the patient through specifically developed semiautomatic software that considered the entry sites to the thorax, the curvature, and the anteroposterior pressure to which the implant would be subjected. The 3D printed template was checked over the patient's chest surface in a fitting ambulatory session to make modifications to the template if necessary. The number of implants was determined based on the processing software recommendations and the anatomy of the sternum (particularly the percentage of sternum that lay behind the anterior costal line) and the prevision was accurate in all patients. The final metallic prebent implants were custom made in according to the 3D printed template and manufactured in surgical steel or titanium. During the surgery the implants were placed on the patient's chest to determine the final location and direction, then the implant tied to a tape was pulled back through a retrosternal tunnel. The removal of implants was planned two–three years later. When the tip of the sternum ended up between two bars resulting in undercorrection, the bars were placed in the crossed position instead of parallel (13.8% of patients). The number of implants introduced was 337, ranging from 1 to 4 per patient (2 or 3 bars in the majority of cases). Of the 130 patients, 120 (92.3%) had an optimal "implant-deformity" match not requiring any modifications; in seven patients (5.4%) the implant needed minimal rebending(achieved without flipping the bars); in two patients (1.5%) the implants were too short; and in one patient (0.8%) with an extremely asymmetric chest the implant was removed and rebent intraoperatively.

Comparing the operative time of patients with pectus excavatum treated in the same hospital before the 3D printing model implementation (group A,) with the cohort described in the study (group B), a significant reduction was observed (group A 125.4 $\pm$ 30.7, vs. group B 87.6 $\pm$ 49.9 min; $p < 0.0001$), also considering the operative time adjusted to the number of bars per operation (group A 78.1 $\pm$ 31.7, vs. group B 41.8 $\pm$ 14.7 min per implant; $p < 0.0001$). The number of bars introduced was increased in the 3D printing group (group A 1.7 $\pm$ 0.6, vs. group B 2.6 $\pm$ 0.5 implants per patient; $p < 0.0001$). There were no intraoperative complications including bleeding. Postoperative complications included two patients with pleural effusion prior to discharge.

Wang et al. [15] (2020) treated six patients with pectus excavatum with the non-thoracoscopic extrapleural Nuss procedure. All patients received a 3D CT scan (1.25 mm slice thickness), and the images were exported as DICOM files to the imaging software to elaborate a 3D thoracic model. A flexible ribcage model was printed in polylactic acid. The flexibility of the materials allowed simulation of the pectus excavatum correction, thus predicting the repair efficacy. The thoracic model was used to study the dimensions of the chest deformity, to plan the trajectory of the steel bars, and to find the optimal substernal force point through simulation of the actual procedure. The repair efficacy was found to be uniform between the prediction on the 3D printed model and the surgical procedure on the patients.

Huang et al. [16] retrospectively analyzed data of 419 patients that underwent the Nuss procedure and selected 357 patients between the traditional Nuss procedure (TN, 342 patients) and the 3D printing model-assisted Nuss procedure (3DPMAN, 15 patients). In 3DPMAN, a standard spiral CT of the thorax was used to generate a 3D-reconstructed thorax model through a processing software. Custom-made template pectus bars and a predicted postoperative 3D thorax model were produced through a computer-aided design process. The template bars were 3D printed with polylactic acid, then metallic pectus bars were manufactured according to a 3D template in the 3DPMAN group and according to a metallic measuring tape in the TN group.

In the TN group, six patients (1.7%) experienced flipping of the metallic bars, two patients experienced migration (0.56%), and two patients (0.56%) experienced dislocation of the metallic bars. No patient experienced surgery-related complications or dislocation of the bars in the 3DPMAN group. A shorter operative time was observed in the 3DPMAN group compared with the TN group (60.36 vs. 74.34 min, $p < 0.001$), fewer pectus bar insertion (1.000 versus 1.360, $p < 0.001$), and better morphological correction (ΔHI: 20.34% versus 10.06%, $p < 0.001$).

Gaspar Pérez et al. [17] prospectively collected data of six patients treated for pectus excavatum with the 3D printed model-assisted Nuss procedure. A diagnostic CT scan was performed in all patients (0.625 mm thickness). The number of bars needed for correction and the appropriate intercostal space for insertion were established preoperatively based on CT scan images and physical characteristics of the patients, then the CT images were uploaded to the segmentation and design software to simulate the morphological repair and establish the Nuss bar size and length. The simulated bar model and chest anatomical model were 3D printed in polylactic acid, then the customized Nuss bar was reproduced in titanium according to the 3D printed bar model. A single Nuss bar was introduced in all cases, and the median operating time was 82 min. There was no need for intraoperative bar replacement or removal, while a single patient experienced intraoperative complications (4 mm distal transfixation perforation of the lingula, with self-limited bleeding and absence of air leaks). No postoperative complications were observed in any patients. Correction was defined as highly satisfactory in all patients.

3.4. Chest Wall Tumor Resection and Reconstruction

Wu et al. [18] reviewed data of six patients with thoracic wall tumors that underwent a 3D printed assisted resection and reconstruction and 10 patients treated with the conventional surgery. The patients in the 3D assisted group underwent thin-slice CT scan, then the DICOM images were imported into the processing software to elaborate a 3D model of the chest wall tumor and adjacent structures. The 3D model, printed using a liquid photosensitive resin, was used to plan the surgery, focusing on the 3D morphology, location, and spatial relationships of the tumor. The titanium implant plate for reconstruction was designed according to the surgical resection line drawn on the 3D printed model, to best fit the thoracic wall defect. Follow-up was conducted at 15 and 90 days after the surgery. The 3D conformal titanium plates were in all cases completely consistent with chest wall defect after tumor resection, even though the tumors of the six patients were different in size and the volume was relatively large. In the 3D printing group, less intraoperative bleeding was reported in relation to the conventional surgery group. No postoperative complications, including plate displacement, were observed in the 3D = assisted group, and there were no significant changes in respiratory function after the surgery ($p < 0.05$). In the conventional surgery group, four patients had complications that consisted of infection and puncture of the artery or skin by the titanium plate. Postoperative pain score was lower in the 3D group. Recovery time was reported to be shortened in 3D group (1 week vs. 2 weeks).

Wang et al. [19] (2019) treated 18 patients that underwent wide resection of a chest wall tumor using 3D printed polyetheretherketone (PEEK) implants. In eight cases the tumor affected the sternum and in 10 cases the ribs. For all patients, a CT scan with slice thickness at 0.90 mm was performed using a 64-detector CT scanner. The images were imported into the processing software for surgical planning and to design 3D PEEK protheses of the ribs or the sternum. The wide excision of the tumor resulted in a defect in the anterior chest wall of at least 8 cm × 8 cm; the reconstruction followed the "sandwich technique". The mean duration of the surgery was 174 ± 54 min, the mean blood loss was 297 ± 235 mL. No complications were observed in the 6 to 12 months after the operation. The respiratory function was tested 1 week before the operation and 3 months after the operation, showing a mean reduction in FVC of 0.39 ± 0.28 L (14.0% of the preoperative FVC value). The postoperative FVC was reduced significantly in both ribs and sternum patients ($p < 0.001$), whereas no significant difference was observed in postoperative MVV and FEV1/FVC.

4. Discussion

In preoperative planning, visualization of the anatomy and pathology of the patient is essential for the surgeon. Compared with standard CT images, 3D reconstruction offers the surgeon the possibility to examine and study the pathology in a more realistic way. Three-dimensional printed models place the 3D reconstruction directly into the surgeon's hands, opening a new phase in surgical planning and training.

The results of our review highlighted some potential benefits of the 3D printing application in chest wall reconstruction. The realization of 3D thorax models contributes to a better understanding of the anatomy of the chest wall defect, and therefore enables the surgeon to plan the surgical steps in advance. An accurate selection of number, location, and direction of implants was made before the operation in all the studies of our review, with the aid of 3D reconstruction and with a 3D printed model of the thorax. Regarding selection of the incision site, Chen et al. [13] and Zhou et al. [12] recorded in their studies a smaller incision site in 3D printing-assisted surgery compared to conventional SSRF. Surgical planning and rehearsal of the procedure can contribute to reducing operative time, as stated by three studies [13,14,16] in our review that compared the 3D-assisted group operating time with that of a conventionally treated control group.

Another interesting aspect is the possibility of manufacturing a custom-made well-fitting prosthesis with minimal need for intraoperative adjustments [12,14–18], often needed in conventional procedures of bar implant or plate insertion [20,21]. Intraoperative adjustments of bars and plates can prolong the operative time and can cause scratching of the implants' surface, which may damage the surrounding tissues and favor adherences. A 3D printed template fitting directly on the patient and simulation of the procedure on the thorax model allowed anticipation of the modifications needed to the implant.

In all studies, no major intraoperative or postoperative complications were found in the 3D printed-assisted surgeries. Wu et al. [18] observed less intraoperative bleeding in the 3D-assisted group compared to the conventional group. The difference between preoperative and postoperative respiratory function was assessed by two studies [18,19] with contrasting results. Wu et al. [18] found no significant difference, whereas Wang et al. [19] observed a reduced FVC and no difference in postoperative MVV and FEV1/FVC.

Huang et al. [16] reported some cases of bars' dislocation and flipping in the traditional surgery group, compared to no cases in the 3D printing-assisted group. Referring to morphological correction, only Huang et al. [16] used a quantitative method to evaluate this outcome, detecting a better correction in patients treated with a 3D printing-assisted technique in comparison to the conventional method, whereas all the other studies recorded satisfactory morphological correction and good implant fitting but did not support this data with a statistical comparison to conventional techniques. Recovery time was directly addressed by Wu et al. [18], who noted a shorter recovery time in the 3D-assisted group and a lower postoperative pain score; and by Zhou et al. [12], who reported a NRS score of 4 or less, and consequently stopped analgesic drugs at 7 days post-procedure in the majority of patients.

For the matter of direct prosthetic printing, Wang et al. [19] interestingly reported the use of sternum and ribs prostheses that were 3D printed in PEEK as an alternative to conventional titanium prostheses due to the physical properties of PEEK, which has a low elastic modulus and similar flexural and tensile strength to the sternum and ribs, and shorter manufacturing time [22,23]. It should be noted that there is not much evidence in the literature regarding the direct implantation of 3D printed prostheses for thoracic wall reconstruction, and the majority of the studies reporting such an application at present are case reports, and were therefore not included in our review.

Production time is a known limitation in 3D printed model applications in surgery, and this is also true of thoracic wall reconstruction. Particularly in emergency situations, such as ribcage fractures associated with damage to the internal organs, time of intervention is critical and the application of 3D printing, which requires at least 5 h for the production of a thoracic model [13], may not be convenient. Conversely, in elective surgery it is possible to schedule the surgical steps in advance and, even if the production time requires several hours, the procedure can be performed without delay. Bellia-Munzon et al. [14] reported 1 working day to print the template and 3 working days for the manufacture of the final metallic bars. Chen et al. [13] reported a minimum of 5–6 h to print the ribcage model. Wang et al. [19] reported approximately 30 h to completely manufacture a PEEK implant.

Another limitation of the application of 3D printing to surgery is the cost, which comprises the printer cost, the processing software cost, and the printing materials cost. The 3D printing utilization cost may be counterbalanced by the shorter operative time [24] and by the benefits that derive from the personalization of the implants, although there is an open debate on this topic. Bellia-Munzon et al. [13] stated that the cost of the implants projected with the assistance of 3D printing is not superior to that of conventional implants. Huang et al. [15] observed that manufacturing a customized prebent bar based on CT images would be more expensive than modeling the bar based on the 3D printing template, and that the costs were further reduced by the fewer insertion of bars in the 3D printing-assisted group.

The main limitation of our review is the reduced number of experimental studies on the topic and the lack of prospective randomized studies. The recent development of 3D printing applications in the surgery field and the fact that not all hospitals are equipped with 3D printers can partly explain the reduced evidence on the subject.

Furthermore, only four of the eight studies presented a control group, which allows us to perform a stronger evaluation of the differences between 3D printing-assisted techniques and conventional techniques. The small sample size in most of the studies is also a crucial limitation, which affects the variability of the data and leads to the risk of selection bias.

5. Conclusions

The application of 3D printing models in chest wall reconstruction can be useful for designing and manufacturing personalized implants, specifically designed on the anatomy of the patient, and therefore reducing intraoperative adjustments. The results of morphological correction and postoperative recovery after 3D printing-assisted surgery were satisfactory in all studies, with a low rate of complications. Although our data point towards a better outcome regarding prosthesis fitting, accuracy of surgical planning, and reduction in operative time in 3D printing-assisted procedures, due to the limitations of our study, more evidence is needed to prove the presented findings.

Author Contributions: Conceptualization, B.L. and A.F.; methodology, A.C. and M.B.; software, A.N.; validation, G.N., G.O. and R.M.; formal analysis, D.P.; investigation, F.C.; resources, V.D.F.; data curation, G.M., F.F., A.C.I. and G.V.; writing—original draft preparation, B.L.; writing—review and editing, M.S. and A.F.; visualization, G.V.; supervision, A.F. and M.S. All authors have read and agreed to the published version of the manuscript.

Funding: This research received no external funding.

Institutional Review Board Statement: Not applicable.

Informed Consent Statement: Not applicable.

Data Availability Statement: The data were available from the articles cited in the review and reported in the References.

Conflicts of Interest: The authors declare no conflict of interest.

References

1. Pascal, A.T.; Laurent, B. Prosthetic reconstruction of the chest wall. *Thorac. Surg. Clin.* **2010**, *20*, 551–558.
2. Sanna, S.; Brandolini, J.; Pardolesi, A.; Argnani, D.; Mengozzi, M.; Dell'Amore, A.; Solli, P. Materials and techniques in chest wall reconstruction: A review. *J. Vis. Surg.* **2017**, *3*, 95. [CrossRef]
3. Goldsmith, I.; Evans, P.L.; Goodrum, H. Chest wall reconstruction with an anatomically designed 3-D printed titanium ribs and hemi-sternum implant. *3D Print. Med.* **2020**, *6*, 26. [CrossRef]
4. Ganguli, A.; Pagan-Diaz, G.J.; Grant, L.; Cvetkovic, C.; Bramlet, M.; Vozenilek, J.; Kesavadas, T.; Bashir, R. 3D printing for preoperative planning and surgical training: A review. *Biomed. Microdevices* **2018**, *20*, 65. [CrossRef]
5. Zhuang, Y.-D.; Zhou, M.-C.; Liu, S.-C.; Wu, J.-F.; Wang, R.; Chen, C.-M. Effectiveness of personalized 3D printed models for patient education in degenerative lumbar disease. *Patient Educ. Couns.* **2019**, *102*, 1875–1881. [CrossRef]
6. Lu, T.; Shao, Z.; Liu, B.; Wu, T. Recent advance in patient-specific 3D printing templates in mandibular reconstruction. *J. Mech. Behav. Biomed. Mater.* **2020**, *106*, 103725. [CrossRef]
7. Hay, J.A.; Smayra, T.; Moussa, R. Customized Polymethylmethacrylate Cranioplasty Implants Using 3-Dimensional Printed Polylactic Acid Molds: Technical Note with 2 Illustrative Cases. *World Neurosurg.* **2017**, *105*, 971–979.
8. Diment, L.E.; Thompson, M.S.; Bergmann, J.H.M. Clinical efficacy and effectiveness of 3D printing: A systematic review. *BMJ Open* **2017**, *7*, e016891. [CrossRef] [PubMed]
9. Gu, B.K.; Choi, D.J.; Park, S.J.; Kim, Y.-J.; Kim, C.-H. 3D Bioprinting Technologies for Tissue Engineering Applications. *Adv. Exp. Med. Biol.* **2018**, *1078*, 15–28.
10. Moher, D.; Shamseer, L.; Clarke, M.; Ghersi, D.; Liberati, A.; Petticrew, M.; Shekelle, P.; Stewart, L.A. Preferred reporting items for systematic review and meta-analysis protocols (PRISMA-P) 2015 statement. *Syst. Rev.* **2015**, *4*, 1. [CrossRef] [PubMed]
11. Downs, S.H.; Black, N. The Feasibility of Creating a Checklist for the Assessment of the Methodological Quality Both of Randomised and Non-Randomised Studies of Health Care Interventions. *J. Epidemiol. Community Health* **1979**, *52*, 377–384. [CrossRef]
12. Zhou, X.; Zhang, D.; Xie, Z.; Yang, Y.; Chen, M.; Liang, Z.; Zhang, G.; Li, S. Application of 3D printing and framework internal fixation technology for high complex rib fractures. *J. Cardiothorac. Surg.* **2021**, *16*, 5. [CrossRef]
13. Chen, Y.-Y.; Lin, K.-H.; Huang, H.-K.; Chang, H.; Lee, S.-C.; Huang, T.-W. The beneficial application of preoperative 3D printing for surgical stabilization of rib fractures. *PLoS ONE* **2018**, *13*, e0204652. [CrossRef]
14. Bellia-Munzon, G.; Martinez, J.; Toselli, L.; Peirano, M.N.; Sanjurjo, D.; Vallee, M.; Martinez-Ferro, M. From bench to bedside: 3D reconstruction and printing as a valuable tool for the chest wall surgeon. *J. Pediatr. Surg.* **2020**, *55*, 2703–2709. [CrossRef]
15. Wang, L.; Guo, T.; Zhang, H.; Yang, S.; Liang, J.; Guo, Y.; Shao, Q.; Cao, T.; Li, X.; Huang, L. Three-dimensional printing flexible models: A novel technique for Nuss procedure planning of pectus excavatum repair. *Ann. Transl. Med.* **2020**, *8*, 110. [CrossRef]

16. Huang, Y.-J.; Lin, K.-H.; Chen, Y.-Y.; Wu, T.-H.; Huang, H.-K.; Chang, H.; Lee, S.-C.; Chen, J.-E.; Huang, T.-W. Feasibility and Clinical Effectiveness of Three-Dimensional Printed Model-Assisted Nuss Procedure. *Ann. Thorac. Surg.* **2019**, *107*, 1089–1096. [CrossRef]
17. Fillat-Gomà, F.; Coderch-Navarro, S.; Monill-Raya, N.; JE, B.A.; Martínez, S.; San Vicente Vela, B.; Jiménez-Arribas, P.; Güizzo, J.R. Initial experience with 3D printing in the use of customized Nuss bars in pectus excavatum surgery. *Cir. Pediatrica* **2021**, *34*, 186–190.
18. Wu, Y.; Chen, N.; Xu, Z.; Zhang, X.; Liu, L.; Wu, C.; Zhang, S.; Song, Y.; Wu, T.; Liu, H.; et al. Application of 3D printing technology to thoracic wall tumor resection and thoracic wall reconstruction. *J. Thorac. Dis.* **2018**, *10*, 6880–6890. [CrossRef] [PubMed]
19. Wang, L.; Huang, L.; Li, X.; Zhong, D.; Li, D.; Cao, T.; Yang, S.; Yan, X.; Zhao, J.; He, J.; et al. Three-Dimensional Printing PEEK Implant: A Novel Choice for the Reconstruction of Chest Wall Defect. *Ann. Thorac. Surg.* **2019**, *107*, 921–928. [CrossRef] [PubMed]
20. Torre, M.; Guerriero, V.; Wong, M.C.Y.; Palo, F.; Lena, F.; Mattioli, G. Complications and trends in minimally invasive repair of pectus excavatum: A large volume, single institution experience. *J. Pediatric Surg.* **2021**, *56*, 1846–1851. [CrossRef] [PubMed]
21. Choi, J.; Kaghazchi, A.; Sun, B.; Woodward, A.; Forrester, J.D. Systematic Review and Meta-Analysis of Hardware Failure in Surgical Stabilization of Rib Fractures: Who, What, When, Where, and Why? *J. Surg. Res.* **2021**, *268*, 190–198. [CrossRef] [PubMed]
22. Wang, L.; Cao, T.; Li, X.; Huang, L. Three-dimensional printing titanium ribs for complex reconstruction after extensive posterolateral chest wall resection in lung cancer. *J. Thorac. Cardiovasc. Surg.* **2016**, *152*, e5–e7. [CrossRef]
23. Aranda, J.L.; Jiménez, M.F.; Rodríguez, M.; Varela, G. Tridimensional titanium-printed custom-made prosthesis for sternocostal reconstruction. *Eur. J. Cardiothorac. Surg.* **2015**, *48*, e92–e94. [CrossRef] [PubMed]
24. Ballard, D.H.; Mills, P.; Duszak, R. Medical 3D Printing Cost-Savings in Orthopedic and Maxillofacial Surgery: Cost Analysis of Operating Room Time Saved with 3D Printed Anatomic Models and Surgical Guides. *Acad Radiol.* **2020**, *27*, 1103–1113. [CrossRef] [PubMed]

Article

Well-Ordered 3D Printed Cu/Pd-Decorated Catalysts for the Methanol Electrooxidation in Alkaline Solutions

Karolina Kołczyk-Siedlecka *, Dawid Kutyła, Katarzyna Skibińska, Anna Jędraczka, Justyna Palczewska-Grela and Piotr Żabiński

Faculty of Non-Ferrous Metals, AGH University of Science and Technology, Mickiewicza 30, 30-059 Krakow, Poland; kutyla@agh.edu.pl (D.K.); kskib@agh.edu.pl (K.S.); kwiec@agh.edu.pl (A.J.); palczews@agh.edu.pl (J.P.-G.); zabinski@agh.edu.pl (P.Ż.)
* Correspondence: kkolczyk@agh.edu.pl

Abstract: In this article, a method for the synthesis of catalysts for methanol electrooxidation based on additive manufacturing and electroless metal deposition is presented. The research work was divided into two parts. Firstly, coatings were obtained on a flat substrate made of light-hardening resin dedicated to 3D printing. Copper was deposited by catalytic metallization. Then, the deposited Cu coatings were modified by palladium through a galvanic displacement process. The catalytic properties of the obtained coatings were analyzed in a solution of 0.1 M NaOH and 1 M methanol. The influence of the deposition time of copper and palladium on the catalytic properties of the coatings was investigated. Based on these results, the optimal parameters for the deposition were determined. In the second part of the research work, 3D prints with a large specific surface were metallized. The elements were covered with a copper layer and modified by palladium, then chronoamperometric curves were determined. The application of the proposed method could allow for the production of elements with good catalytic properties, complex geometry with a large specific surface area, small volume and low weight.

Keywords: electroless metallization; catalysts; 3D printing

Citation: Kołczyk-Siedlecka, K.; Kutyła, D.; Skibińska, K.; Jędraczka, A.; Palczewska-Grela, J.; Żabiński, P. Well-Ordered 3D Printed Cu/Pd-Decorated Catalysts for the Methanol Electrooxidation in Alkaline Solutions. *Technologies* **2021**, *9*, 6. https://doi.org/10.3390/technologies9010006

Received: 27 November 2020
Accepted: 5 January 2021
Published: 8 January 2021

Publisher's Note: MDPI stays neutral with regard to jurisdictional claims in published maps and institutional affiliations.

1. Introduction

Additive manufacturing is a technique that is increasingly used in new technologies [1–3]. The development of the digital revolution and 3D printing methods are changing the methods used for the production of functional objects. The use of 3D printing techniques allows for a quick transition from a digital model to a physical object. This ensures great flexibility in adapting a given geometry, as opposed to classical production methods like machining or casting for metallic parts or injection molding for plastics [4]. The advantage of this technique is the possibility of producing elements from materials such as plastic, metal or materials based on composites [5–7]. In addition, the modification of elements produced by the 3D printing method is also carried out, for example by covering them with metallic coatings. Thanks to the listed advantages and many possibilities, additive manufacturing has found many applications in many fields, for example in new technologies, medicine [8], catalysis [9–11] and electrochemistry [12–17].

Direct methanol fuel cells (DMFCs) are a promising solution for the problems of energy conversion. They are characterized by their small size, high energy conversion efficiency, low working temperature and the availability of methanol as fuel [18]. However, there are some limitations to introducing these cells into commercial markets. This is related to the low efficiency of the anode reaction due to the slow kinetics of methanol electrooxidation and the destruction of the electrode surfaces [19–21].

Good catalytic properties characterize noble metals such as Pt [22–24], Pd [25–27] and Ag [28–30]. However, these metals are expensive, and this limits their application possibilities.

Copper is not a very popular catalyst for the electrooxidation process due to its high sensitivity and its reactivity towards oxygen. Nevertheless, there are research works focused on the application of this metal or its alloys. The catalytic properties of other metals, such as nickel [31], copper [32,33] and their alloys Ni-Cu [34,35] have been analyzed. Catalysts synthesized with noble metals, e.g., Pd-Ni [36–39], Pt-Ni [22,40], Pd-Cu [26,41–43], Pt-Cu [44–47], have also been tested.

In the literature, catalytic coatings have often been deposited on conductive substrates such metals of graphite. Unfortunately, these materials often have a high density or they are more challenging to manufacture, especially with high-precision techniques. One of the solutions is the combination of additive manufacturing in terms of the possibility of obtaining complex geometry inside an element and the process of electroless deposition of metallic coatings [48]. In this paper, the results of research related to the synthesis of Cu/Pd coatings by electroless deposition on a resin substrate dedicated to 3D printing are presented. A novelty in this work is the development of electroless deposition on a complex surface. In our previous work [49,50], metallic electroless coatings were obtained on a flat surface. The obtained materials were analyzed in terms of catalytic properties in the methanol electrooxidation reaction. The presented experimental works are intended to develop a new type of catalysts.

2. Materials and Methods

In the experimental work, chemicals with analytical purity were used (ChemLand company, Poland). The main stage of research was the metallization of light-hardened resins. The substrate was made of FormLabs Form 2 (CadXpert company, Poland) resins (Clear and Gray) dedicated to the stereolithography (SLA) method. Both of them are characterized by chemical resistance, as described by the manufacturer [51,52].

A liquid resin with a volume of 1 mL was dropped onto a glass substrate, and then cured using a UV lamp (power 48 W) for 1 min. The cured samples were washed in two stages in isopropanol for 20 min every step, according to the manufacturer's instructions. Then, they were washed with demineralized water and dried.

The cured samples were degreased in 5 wt% NaOH solution at 70 °C for 10 min to remove other impurities from the surface. Then, the samples were washed in demineralized water and etched for 1 min at 70 °C in a chromic acid solution prepared as follows: 50 g Cr_2O_3, 1500 g H_2SO_4, 250 g H_2O. The chromium(IV) ions were neutralized and removed from the surface in 5 wt% HCl and 5 wt% $K_2S_2O_5$ solution for 3 min at room temperature. Then, the samples were washed in demineralized water. The surface of the resin was activated by Pd(II) ions, the samples were placed into 1 g/L $PdCl_2$ solution for 30 min, and during this time the ions were adsorbed on resin surface. After this step, the samples were placed into freshly prepared 20 g/L $NaBH_4$ solution to reduce the palladium ions.

After activation and reduction, the samples were thoroughly washed with demineralized water and placed into the metallization solution: 10 g/L $CuSO_4 \cdot 5H_2O$, 10 g/L tartaric acid, 10 mL/L formalin. The pH of the solution was modified by NaOH, and it was equal to 12. The temperature of solution was 40 °C. The copper was deposited for 10 and 20 min. Then, the samples were washed using demineralized water and placed for galvanic displacement into 2 g/L $PdCl_2$ solution for 30 s to 5 min. To compare the parameters of electroless deposited coatings with bulk material, the copper sheets were placed in the same solution for 1 and 5 min.

All samples were tested electrochemically in 0.1 M NaOH + 1 M methanol solution in a three-electrode system at room temperature. The first electrode was a Cu/Pd sample, the counter electrode was a Pt sheet, and as the reference electrode the saturated calomel electrode (SCE) was used. Cyclic voltammetry experiments were performed in the potential range between hydrogen and oxygen evolution with different scan rates.

Using SolidWorks 2018 SP 4.0 software and the FormLabs Form 2 3D printer, cylinder-shaped elements with a diameter of 2 cm and a height of 3 cm were designed and printed. The elements were characterized by large specific surface in relation to the dimensions.

Similar to the previous samples, the elements were metallized by copper and decorated by palladium. Cyclic voltammogram measurements and a chronoamperometric curve were performed to evaluate catalytic properties. As was carried out previously, the elements were tested in 0.1 M NaOH + 1 M CH3OH solutions at room temperature in the three-electrode system. The working electrode was the metallized element and the counter electrode was a Pt sheet. The cycling voltammograms were detected in the range from −1.0 to 1.5 V vs. SCE.

The obtained coatings were analyzed using scanning electron microscopy (SEM) with a JEOL—6000 Plus. The chemical composition and distribution of elements were determined using energy dispersive X-ray spectroscopy (EDS) analysis. The catalytical tests were performed using a BioLogic SP-200 potentiostat.

3. Results and Discussion

3.1. Physical Characterization

In this work, the concentration of copper and palladium in the obtained coatings was determined. For the selected sample, the analysis of Cu and Pd distribution was performed. In the main part of the research work, a series of catalytic tests for the methanol oxidation reaction were performed. SEM pictures of the coatings before and after catalytic tests were taken.

The elemental composition of the coatings obtained as a result of electroless (EL) copper deposition and galvanic displacement by palladium is shown in Figure 1. The copper concentration is marked in blue and the palladium concentration in red. The different types of obtained copper samples are marked with different symbol shapes. The change in Pd concentration was linear with the time of galvanic displacement. The palladium concentration was not dependent on the Cu deposition time. In both cases of 10 min and 20 min of electroless Cu metallization, the Pd content ranged from about 2.75% (30 s) to 16.5% by mass (5 min). In the case of Pd galvanic displacement on metallic copper, the palladium content after 5 min was equal to 15.12% by mass and on the Cu coatings this value was approx. 16.8% and 16.6% on the copper deposited for 10 and 20 min, respectively.

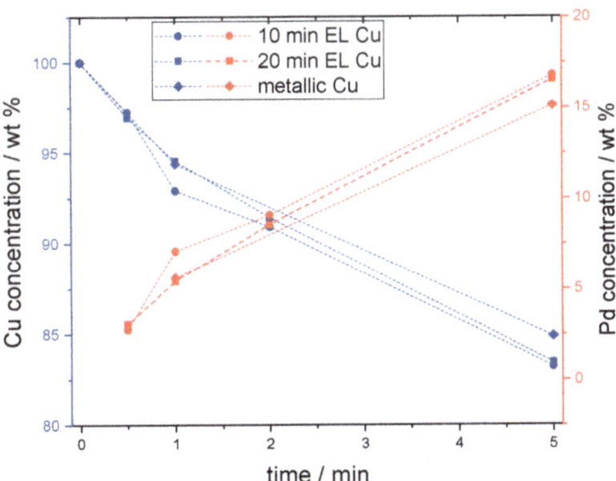

Figure 1. Composition of coatings depending on galvanic displacement time.

The SEM image taken for the coating obtained after 10 min of electroless Cu metallization and 5 min of galvanic exchange of palladium is presented in Figure 2. It can be seen that a smooth surface was obtained, and a slight precipitation can be observed. The mapping analysis showed that the distribution of Pd was homogeneous.

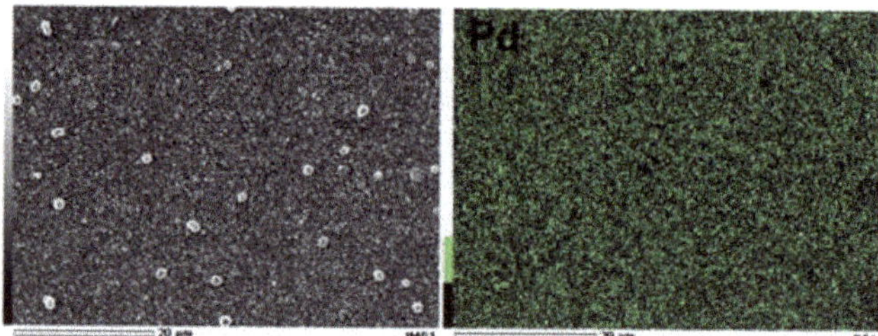

Figure 2. Mapping analysis of Cu/Pd after 10 min of electroless deposition of Cu and 5 min of galvanic displacement Pd. Magnification: ×2000.

3.2. Electrooxidation of Methanol on the Cu/Pd-Decorated Coatings

The methanol electrooxidation activity of Cu electrolessly deposited coatings modified by palladium was investigated in 0.1 M NaOH + 1 M CH$_3$OH solutions at room temperature using electrochemical techniques. Figure 3 presents cyclic voltammograms of materials, depending on electroless deposition of Cu and Pd galvanic displacement time (Figure 3a,b). As shown the presented cyclic voltammograms, a change in the anodic peak connected with the oxidation of methanol was observed. Depending on the time of Pd deposition by galvanic displacement and thus the different concentration, the maximum potential value changed, ranging from 0.8 V vs. SCE for 5 min Pd (about 16 wt. %) to 1.2 V vs. SCE after 2 min of Pd (about 10 wt. %) on the Cu substrate deposited for 10 min (Figure 3a). In the case of the Cu coatings deposited for 20 min, the anode peaks varied to a lesser extent, from 1.2 to 1.36 V vs. SCE (Figure 3b), with the highest activity observed for the modification by Pd for 30 s. To compare the catalytic properties, the diagrams for materials obtained on metallic copper are also presented (Figure 3c). Experiments were carried out on a copper substrate modified by palladium for 1 min and 5 min. No significant difference was observed between the measurements carried out; the potential of the anode peak is 1.3 V vs. SCE (Figure 3c).

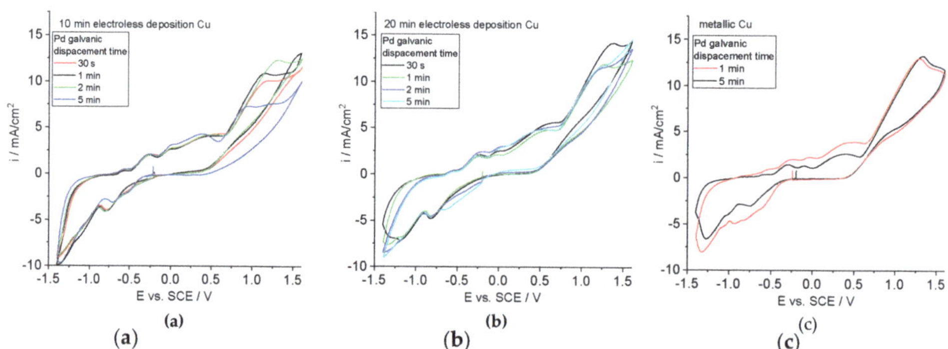

Figure 3. Cyclic voltammograms of Cu/Pd-decorated coatings depending on the substrate type and Pd galvanic displacement time. The substrates were as follows: 10 min (**a**), 20 min (**b**) electrolessly deposition copper and metallic copper (**c**). The measurements were performed in 0.1 M NaOH + 1 M methanol solution, scan rate: 50 mV/s.

The cyclic voltammogram curves for different scan rates were determined. Figure 4 shows them for two exemplary samples—the coatings obtained in 20 min Cu electroless metallization + 30 s and 5 min Pd galvanic displacement. As the scan rate increases, the anode peak from the oxidation of methanol increases. The same tendency was shown in all the analyses. Cyclic voltammograms of the remaining samples depending on the scan rates are presented in the Supplementary Materials (Figures S1–S6).

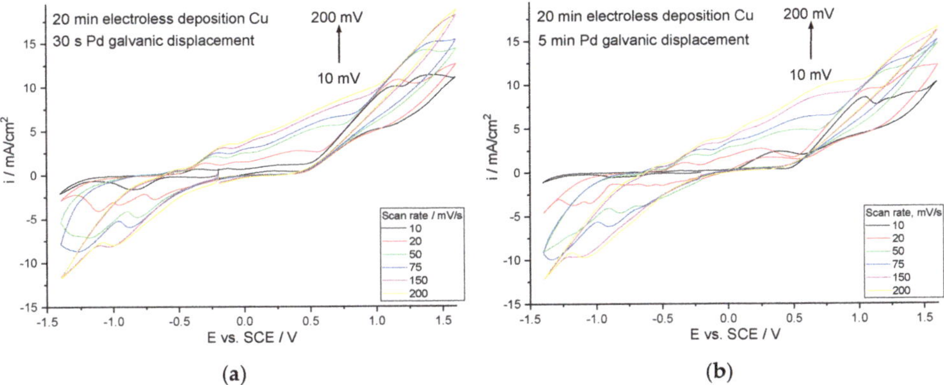

Figure 4. Cyclic voltammograms of obtained coatings in 0.1 M NaOH + 1 M methanol solution at different scan rates. The deposition parameters: 20 min electroless deposition of Cu, modification by palladium for 30 s (**a**) and 5 min (**b**).

Based on the cyclic voltammogram curves presented in Figure 4 and in the Supplementary Materials (Figures S5 and S6), the change in the relationship between the current density and the square root of the scan rate was determined (Figure 5a). This relationship has been shown to be linear, therefore it can be concluded that the methanol electrooxidation reaction takes place through diffusion control [44]. For all cases, a high correlation coefficient R^2 was obtained. Figure 5b shows that the potential changed linearly with the logarithm of the scan rate, suggesting that methanol oxidation on the surfaces is an irreversible process [44].

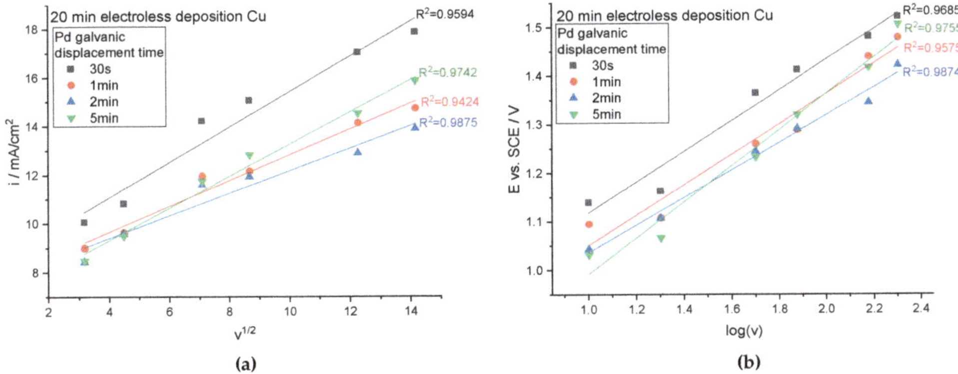

Figure 5. Characteristics of the coatings depending on the Pd galvanic displacement time. Change in anodic peak current with the square roots of scan rate (**a**), change in anodic peak potential with logarithm of scan rate (**b**). The copper was deposited for 20 min and modified by Pd.

The SEM pictures of the coatings prepared for electrochemical analysis and after catalytic tests are presented in Figure 6. The Cu coatings obtained electrolessly were characterized by continuity. In the case of copper deposited for 10 min, small holes were observed. Small irregularly shaped or spherical precipitates were observed on the coating surfaces. Precipitations came from the electroless copper deposition process, and not from Pd galvanic displacement, which can be observed in the mapping analysis (Figure 2) and in comparison to the Cu coatings obtained after 20 min (Figure 6). In the case of Cu deposited for 20 min, the coatings had a homogeneous morphology and small crystalline coatings were observed. After catalytic tests, in the case of Cu/Pd materials obtained after 10 min of electroless copper deposition, the destruction of the coatings was clearly visible.

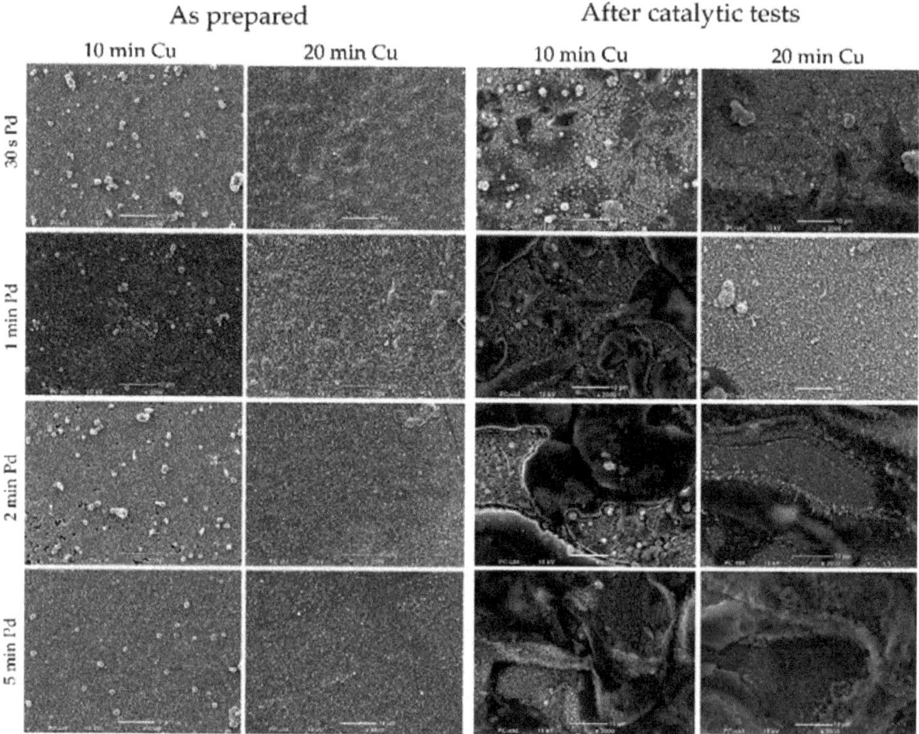

Figure 6. SEM pictures of electroless-deposited Cu modified by Pd, depending on copper deposition and galvanic displacement time as prepared and after electrochemical tests, magnification ×2000.

In some cases (e.g., after 30 s of Pd galvanic displacement) characteristic precipitations were still visible. Coatings based on Cu deposited for 20 min were characterized by less damage of the coatings. After the tests, the coatings were still stable and showed good adhesion. The bluish discoloration of the precipitate was clearly visible, indicating copper oxidation.

To assess the quality of coatings prepared on metallic copper, the SEM pictures were taken (Figure 7). Slight cracks were visible on the surface, and slight irregular precipitations were visible after 5 min of palladium deposition. Surface destruction was observed after catalytic tests—in the case of the coating deposited after 1 min with Pd galvanic displacement, the size of cracks on the surface increased. In the case of a higher palladium content, most of the precipitates from the surface disappeared. As with the resin-deposited coatings, it was observed that the coatings were oxidized after the tests.

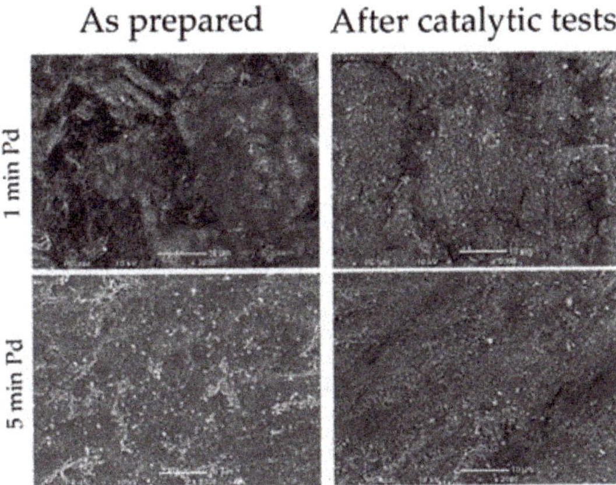

Figure 7. SEM pictures of metallic Cu modified by Pd, depending on galvanic displacement time as prepared and after electrochemical tests, magnification ×2000.

3.3. Methanol Electrooxidation on 3D Printed Elements

A prototype of a catalyst for the electrooxidation of methanol in basic solutions was made, and its design is presented in Figure 8. The cylindrical prints were designed based on common geometry [11]. Round holes were made parallel to and perpendicular to the cylinder's axis.

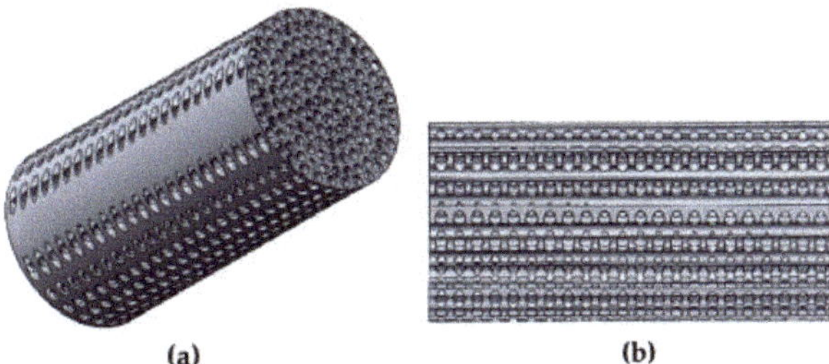

Figure 8. Drawing (a) and cross-section (b) of the catalyst made in the SolidWorks software.

Figure 8 shows the digital design of the element. The cross-section shown in Figure 8b shows that the element is characterized by high porosity, thanks to which it was possible to obtain a large specific surface, which is equal to 230 cm^2.

Two elements were prepared and metallized by copper according to the procedure described previously; the Cu metallization time was equal to 20 min. Then, the metallized elements were placed into 2 g/L PdCl$_2$ solutions for 2 and 5 min. The concentration of palladium in Cu/Pd coatings was determined by EDS.

The anodic peak was determined from the cyclic voltammogram curves presented in Figure 9a. The obtained potential was chosen for potentiostatic measurements, which were performed over 15 min in the same solutions. In case of the coatings modified by Pd for 5 min, the maximum determined potential was equal to 1.3 V vs. SCE. An anode potential of 1.05 V vs. SCE was observed on the curve of the coating deposited for 20 min with Cu and modified for 2 min by Pd. For the chronoamperometry curves, the stability of the current over time was observed (Figure 9b). After a certain period, both curves tended to plateau, when the Pd concentration was near 20% and over 10% by mass.

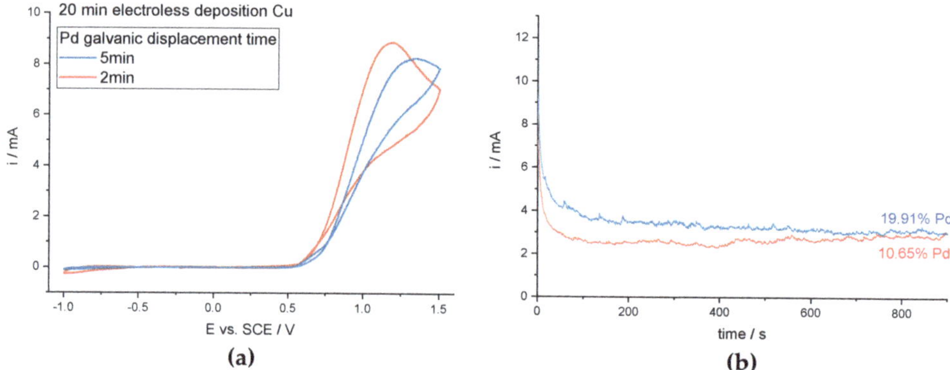

Figure 9. Cyclic voltammograms (**a**) and chronoamperometry (**b**) of metallized 3D prints. The Cu substrate was deposited in 20 min and modified by Pd in galvanic displacement in 2 and 5 min.

Figure 10 presents SEM pictures of Cu/Pd coatings obtained on 3D prints, depending on the modification time with palladium, as prepared and after catalytic tests. The coatings were characterized by a similar morphology, regardless of the palladium galvanic displacement time. Fewer minor precipitates were also observed for the shorter time of Pd deposition. The coatings did not deteriorate after catalytic tests; they remained compact and adhered to the substrate. Fewer precipitates on the surface were observed.

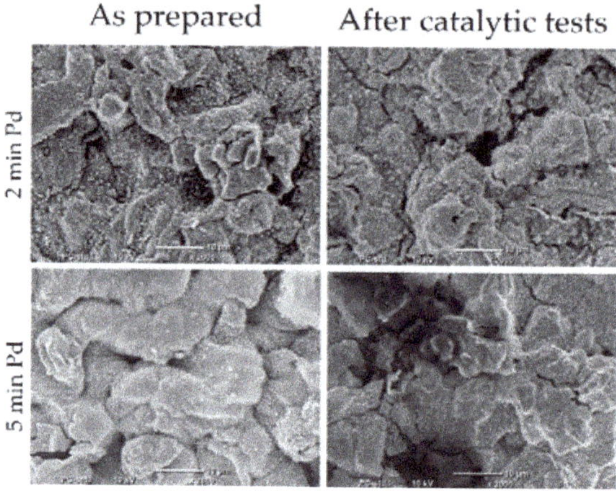

Figure 10. SEM pictures of coatings deposited on 3D prints as prepared and after catalytic tests, magnification ×2000.

4. Conclusions

The aim of this work was to develop a prototype of a catalyst intended for direct methanol fuel cells. A catalyst synthesis based on 3D printing and electroless copper deposition was proposed.

Metallic Cu coatings were modified by palladium through the galvanic displacement process. In the first stage, Cu/Pd coatings on a flat substrate were synthesized. The obtained coatings were characterized by a homogeneous distribution of elements on their surface. Cyclic voltammogram curves were obtained to determine the catalytic properties of Cu/Pd coatings. The coatings formed on the Cu substrate deposited for 20 min were characterized by smaller anode peak potential differences, depending on composition. By determining the linear relationship of currents in the anode peak as a function of the element from the scan rate and the potential from log (v), it was determined that the reaction was diffusion-controlled and irreversible. Based on the SEM pictures, it was determined that coatings obtained on electroless copper deposition for 10 min are porous and unstable—after catalytic tests, the destruction of the coatings was visible. Coatings based on 20 min EL Cu were less damaged.

On the basis of the results obtained, the metallization parameters of substrates constituting 3D prints with a large specific surface were determined. Composition, cyclic voltammograms and chronoamperometry curves were determined and SEM pictures of the coatings were taken before and after the electrochemical tests. It was found that the coatings were not significantly damaged by the applied potential. Using these techniques, a 3D printed catalyst model was pre-developed.

Supplementary Materials: The following are available online at https://www.mdpi.com/2227-708 0/9/1/6/s1, Figure S1: Cyclic voltammograms of 10 min electroless (EL) copper + 30 s Pd in 0.1 M NaOH + 1 M methanol solution at different scan rates; Figure S2: Cyclic voltammograms of 10 min EL copper + 1 min Pd in 0.1 M NaOH + 1 M methanol solution at different scan rates; Figure S3: Cyclic voltammograms of 10 min EL copper + 2 min Pd in 0.1 M NaOH + 1 M methanol solution at different scan rates; Figure S4: Cyclic voltammograms of 10 min EL copper + 5 min Pd in 0.1 M NaOH + 1 M methanol solution at different scan rates; Figure S5: Cyclic voltammograms of 20 min EL copper + 1 min Pd in 0.1 M NaOH + 1 M methanol solution at different scan rates; Figure S6: Cyclic voltammograms of 20 min EL copper + 2 min Pd in 0.1 M NaOH + 1 M methanol solution at different scan rates.

Author Contributions: Conceptualization, K.K.-S. and D.K.; methodology, K.K.-S. and A.J.; software, K.S.; validation, K.K.-S., P.Ż. and J.P.-G.; resources, K.K.-S.; data curation, K.K.-S.; writing—original draft preparation, K.K.-S.; writing—review and editing, D.K. and J.P.-G.; supervision, P.Ż. All authors have read and agreed to the published version of the manuscript.

Funding: This research was funded by Polish National Science Centre, grant number No UMO-2017/25/N/ST8/01721. The authors are grateful to Faculty of Non-Ferrous Metals for providing space and materials for research.

Institutional Review Board Statement: Not applicable.

Informed Consent Statement: Not applicable.

Data Availability Statement: Data is contained within the article or supplementary material.

Conflicts of Interest: The authors declare no conflict of interest. The funders had no role in the design of the study; in the collection, analyses or interpretation of data; in the writing of the manuscript, or in the decision to publish the results.

References

1. Zhang, F.; Wei, M.; Viswanathan, V.V.; Swart, B.; Shao, Y.; Wu, G.; Zhou, C. 3D printing technologies for electrochemical energy storage. *Nano Energy* **2017**, *40*, 418–431. [CrossRef]
2. Jiménez, M.; Romero, L.; Dom, I.A.; Espinosa, M.D.M.; Domínguez, M. Additive Manufacturing Technologies: An Overview about 3D Printing Methods and Future Prospects. *Complexity* **2019**, *2019*, 9656938. [CrossRef]
3. Yap, Y.L.; Sing, S.L.; Yeong, W.Y. A review of 3D printing processes and materials for soft robotics. *Rapid Prototyp. J.* **2020**, *26/8*, 1345–1361. [CrossRef]
4. Attaran, M. The rise of 3-D printing: The advantages of additive manufacturing over traditional manufacturing. *Bus. Horiz.* **2017**, *60*, 677–688. [CrossRef]
5. Ngo, T.D.; Kashani, A.; Imbalzano, G.; Nguyen, K.T.Q.; Hui, D. Additive manufacturing (3D printing): A review of materials, methods, applications and challenges. *Compos. Part B* **2018**, *143*, 172–196. [CrossRef]
6. Bourell, D.; Pierre, J.; Leu, M.; Levy, G.; Rosen, D.; Beese, A.M.; Clare, A. Materials for additive manufacturing. *CIRP Ann. Manuf. Technol.* **2017**, *66*, 659–681. [CrossRef]
7. Ma, C.; He, C.; Wang, W.; Yao, X.; Yan, L.; Hou, F. Metal-doped polymer-derived SiOC composites with inorganic metal salt as the metal source by digital light processing 3D printing. *Virtual Phys. Prototyp.* **2020**, *15*, 294–306. [CrossRef]
8. Javaid, M.; Haleem, A.; Javaid, M.; Haleem, A. Additive manufacturing applications in medical cases: A literature based review. *Alex. J. Med.* **2019**, *54*, 411–422. [CrossRef]

9. Lind, A.; Vistad, Ø.; Sunding, M.F.; Andreassen, K.A.; Ha, J.; Grande, C.A. Multi-purpose structured catalysts designed and manufactured by 3D printing. *Mater. Des.* **2020**, *187*, 1–8. [CrossRef]
10. Parra-cabrera, C.; Achille, C.; Kuhn, S.; Ameloot, R.; Parra-cabrera, C. 3D printing in chemical engineering and catalytic technology: Structured catalysts, mixers and reactors. *Chem. Soc. Rev.* **2017**, *47*, 209–230. [CrossRef]
11. Bogdan, E.; Michorczyk, P. 3D Printing in Heterogeneous Catalysis—The State of the Art. *Materials* **2020**, *13*, 4534. [CrossRef] [PubMed]
12. Rocha, D.P.; Oliveira, P.R.; Janegitz, B.C.; Bonacin, J.A.; Richter, E.M.; Munoz, R.A.A. Additive-manufactured (3D-printed) electrochemical sensors: A critical review. *Anal. Chim. Acta* **2020**, *1118*, 73–91. [CrossRef]
13. Hong, E.; Ho, Z.; Ambrosi, A.; Pumera, M. Additive manufacturing of electrochemical interfaces: Simultaneous detection of biomarkers. *Appl. Mater. Today* **2018**, *12*, 43–50. [CrossRef]
14. Cheng, T.S.; Zafir, M.; Nasir, M.; Ambrosi, A.; Pumera, M. 3D-printed metal electrodes for electrochemical detection of phenols. *Appl. Mater. Today* **2017**, *9*, 212–219. [CrossRef]
15. Lee, C.; Taylor, A.C.; Nattestad, A.; Beirne, S.; Wallace, G.G. 3D Printing for Electrocatalytic Applications. *Joule* **2019**, *3*, 1835–1849. [CrossRef]
16. Tubío, C.R.; Azuaje, J.; Escalante, L.; Coelho, A.; Guitián, F.; Sotelo, E.; Gil, A. 3D printing of a heterogeneous copper-based catalyst. *J. Catal.* **2016**, *334*, 110–115. [CrossRef]
17. Zhu, C.; Liu, T.; Qian, F.; Chen, W.; Chandrasekaran, S.; Yao, B.; Song, Y.; Duoss, E.B.; Kuntz, J.D.; Spadaccini, C.M.; et al. 3D printed functional nanomaterials for electrochemical energy storage. *Nano Today* **2017**, *15*, 107–120. [CrossRef]
18. Yuda, A.; Ashok, A.; Kumar, A. A comprehensive and critical review on recent progress in anode catalyst for methanol oxidation reaction. *Catal. Rev.* **2020**. [CrossRef]
19. Jorissen, L.; Gogel, V. Direct Methanol: Overview. In *Encyclopedia of Electrochemical Power Sources*; Elsevier: London, UK, 2009; pp. 370–380.
20. Samimi, F.; Rahimpour, M.R. Direct Methanol Fuel Cell. In *Methanol*; Elsevier B.V.: Amsterdam, The Netherlands, 2018; pp. 381–397; ISBN 9780444639035.
21. Liu, H.; Song, C.; Zhang, L.; Zhang, J.; Wang, H.; Wilkinson, D.P. A review of anode catalysis in the direct methanol fuel cell. *J. Power Sources* **2006**, *155*, 95–110. [CrossRef]
22. Nagashree, K.L.; Raviraj, N.H.; Ahmed, M.F. Carbon paste electrodes modified by Pt and Pt–Ni microparticles dispersed in polyindole film for electrocatalytic oxidation of methanol. *Electrochim. Acta* **2010**, *55*, 2629–2635. [CrossRef]
23. Al-saidi, W.S.; Jahdaly, B.A.A.L.; Awad, M.I. Methanol electro-oxidation at platinum electrod: In situ recovery of CO poisoned platinum electrode. *Int. J. Electrochem. Sci.* **2020**, *15*, 1830–1839. [CrossRef]
24. Ejigu, A.; Johnson, L.; Licence, P.; Walsh, D.A. Electrocatalytic oxidation of methanol and carbon monoxide at platinum in protic ionic liquids. *Electrochem. Commun.* **2012**, *23*, 122–124. [CrossRef]
25. Carlos, J.; Gómez, C.; Moliner, R.; Lázaro, M.J. Palladium-Based Catalysts as Electrodes for Direct Methanol Fuel Cells: A Last Ten Years Review. *Catalysts* **2016**, *6*, 130. [CrossRef]
26. Yan, B.; Xu, H.; Zhang, K.; Li, S.; Wang, J.; Shi, Y.; Du, Y. Cu assisted synthesis of self-supported PdCu alloy nanowires with enhanced performances toward ethylene glycol electrooxidation. *Appl. Surf. Sci.* **2018**, *434*, 701–710. [CrossRef]
27. Roy, S.; Mukherjee, P. Palladium and palladium-copper alloy nano particles as superior catalyst for electrochemical oxidation of methanol for fuel cell applications. *Int. J. Hydrogen Energy* **2016**, *41*, 17072–17083. [CrossRef]
28. Shivakumar, M.S.; Krishnamurthy, G.; Ravikumar, C.R.; Bhatt, A.S. Decoration of silver nanoparticles on activated graphite substrate and their electrocatalytic activity for methanol oxidation. *J. Sci. Adv. Mater. Devices* **2019**, *4*, 290–298. [CrossRef]
29. Rinco, A.; Gutie, C. Electrooxidation of methanol on silver in alkaline medium. *J. Electroanal. Chem.* **2000**, *495*, 71–78.
30. Nocun, M.; Jabłon, M. Silver–Alumina Catalysts for Low-Temperature Methanol Incineration. *Catal. Lett.* **2016**, *146*, 937–944. [CrossRef]
31. Salci, A.; Altunbas, E.; Solmaz, R. Methanol electrooxidation at nickel-modified rhodanine self assembled monolayer films: A new class of multilayer electrocatalyst. *Int. J. Hydrog. Energy* **2019**, *44*, 14228–14234. [CrossRef]
32. Heli, H.; Jafarian, M.; Mahjani, M.G.; Gobal, F. Electro-oxidation of methanol on copper in alkaline solution. *Electrochim. Acta* **2004**, *49*, 4999–5006. [CrossRef]
33. Chen, G.; Pan, Y.; Lu, T.; Wang, N.; Li, X. Highly catalytical performance of nanoporous copper for electro-oxidation of methanol in alkaline media. *Mater. Chem. Phys.* **2018**, *218*, 108–115. [CrossRef]
34. Danaee, I.; Jafarian, M.; Forouzandeh, F.; Gobal, F.; Mahjani, M.G. Electrocatalytic oxidation of methanol on Ni and NiCu alloy modified glassy carbon electrode. *Int. J. Hydrog. Energy* **2008**, *33*, 4367–4376. [CrossRef]
35. Mao, Y.; Chen, C.; Fu, J.; Lai, T.; Lu, F.; Tsai, Y. Electrodeposition of nickel—Copper on titanium nitride for methanol electrooxidation. *Surf. Coat. Technol.* **2018**, *350*, 949–953. [CrossRef]
36. Kamyabi, M.A.; Jadali, S. A sponge like Pd arrays on Ni foam substrate: Highly active non-platinum electrocatalyst for methanol oxidation in alkaline media. *Mater. Chem. Phys.* **2021**, *257*, 123626. [CrossRef]
37. Qiu, C.; Shang, R.; Xie, Y.; Bu, Y.; Li, C.; Ma, H. Electrocatalytic activity of bimetallic Pd–Ni thin films towards the oxidation of methanol and ethanol. *Mater. Chem. Phys.* **2010**, *120*, 323–330. [CrossRef]
38. Carvalho, L.L.; Colmati, F.; Tanaka, A.A. Nickel-palladium electrocatalysts for methanol, ethanol, and glycerol oxidation reactions. *Int. J. Hydrogen Energy* **2017**, *42*, 16118–16126. [CrossRef]

39. Maity, S.; Harish, S.; Eswaramoorthy, M. Controlled galvanic replacement of Ni in Ni(OH)2 by Pd: A method to quantify metallic Ni and to synthesize bimetallic catalysts for methanol oxidation. *Mater. Chem. Phys.* **2019**, *221*, 377–381. [CrossRef]
40. Habibi, B.; Dadashpour, E. Carbon-ceramic supported bimetallic Pt e Ni nanoparticles as an electrocatalyst for electrooxidation of methanol and ethanol in acidic media. *Int. J. Hydrogen Energy* **2012**, *38*, 5425–5434. [CrossRef]
41. Yin, Z.; Zhou, W.; Gao, Y.; Ma, D.; Kiely, C.J. Supported Pd–Cu Bimetallic Nanoparticles That Have High Activity for the Electrochemical Oxidation of Methanol. *Chem. Eur. J.* **2012**, *18*, 4887–4893. [CrossRef]
42. Na, H.; Zhang, L.; Qiu, H.; Wu, T.; Chen, M.; Yang, N.; Li, L.; Xing, F.; Gao, J. A two step method to synthesize palladium e copper nanoparticles on reduced graphene oxide and their extremely high electrocatalytic activity for the electrooxidation of methanol and ethanol. *J. Power Sources* **2015**, *288*, 160–167. [CrossRef]
43. Zhang, Z.; Zhang, C.; Sun, J.; Kou, T.; Zhao, C. Ultrafine nanoporous Cu–Pd alloys with superior catalytic activities towards electro-oxidation of methanol and ethanol in alkaline media. *RSC Adv.* **2012**, *2*, 11820–11828. [CrossRef]
44. Poochai, C.; Veerasai, W.; Somsook, E.; Dangtip, S. The influence of copper in dealloyed binary platinum-copper electrocatalysts on methanol electroxidation catalytic activities. *Mater. Chem. Phys.* **2015**, *163*, 317–330. [CrossRef]
45. Poochai, C. Highly active dealloyed Cu@Pt core-shell electrocatalyst towards 2-propanol electrooxidation in acidic solution. *Appl. Surf. Sci.* **2017**, *396*, 1793–1801. [CrossRef]
46. Long, X.; Yin, P.; Lei, T.; Wang, K.; Zhan, Z. Methanol electro-oxidation on Cu@Pt/C core-shell catalyst derived from Cu-MOF. *Appl. Catal. B Environ.* **2020**, *260*, 118187. [CrossRef]
47. Lee, J.; Han, S.; Kwak, D.; Kim, M.; Lee, S.; Park, J.; Choi, I.; Park, H.; Park, K. Porous Cu-rich@Cu3Pt alloy catalyst with a low Pt loading for enhanced electrocatalytic reactions. *J. Alloys Compd.* **2017**, *691*, 26–33. [CrossRef]
48. Bernasconi, R.; Credi, C.; Tironi, M.; Levi, M.; Magagnin, L. Electroless Metallization of Stereolithographic Photocurable Resins for 3D Printing of Functional Microdevices. *J. Electrochem. Soc.* **2017**, *164*, B3059–B3066. [CrossRef]
49. Kolczyk, K.; Zborowski, W.; Kutyla, D.; Kwiecinska, A.; Kowalik, R.; Zabinski, P. Investigation of Two-Step Metallization Process of Plastic 3d Prints Fabricated by SLA Method. *Arch. Metall. Mater.* **2018**, *63*, 1031–1036. [CrossRef]
50. Kołczyk-Siedlecka, K.; Skibińska, K.; Kutyła, D.; Kwiecińska, A.; Kowalik, R.; Żabiński, P. Influence of Magnetic Field on Electroless Metallization of 3d Prints by Copper and Nickel. *Arch. Metall. Mater.* **2019**, *64*, 17–22. [CrossRef]
51. KMK Regulatory Services Inc. Clear Photoreactive Resin for Formlabs 3D Printers Karta Charakterystyki. Available online: https://sklep.3dl.tech/wp-content/uploads/2017/12/Clear_Formlabs-SDS-EU-Polish.pdf (accessed on 15 June 2019).
52. KMK Regulatory Services Inc. Grey Photoreactive Resin for Formlabs 3D Printers Karta Charakterystyki. Available online: https://sklep.3dl.tech/wp-content/uploads/2017/12/Grey_Formlabs-SDS-EU-Polish.pdf (accessed on 15 June 2019).

Article

Metallization of Thermoplastic Polymers and Composites 3D Printed by Fused Filament Fabrication

Alessia Romani [1,2], Andrea Mantelli [1], Paolo Tralli [3], Stefano Turri [1], Marinella Levi [1] and Raffaella Suriano [1,*]

1   Department of Chemistry, Materials and Chemical Engineering "Giulio Natta", Politecnico di Milano, Piazza Leonardo da Vinci 32, 20133 Milano, Italy; alessia.romani@polimi.it (A.R.); andrea.mantelli@polimi.it (A.M.); stefano.turri@polimi.it (S.T.); marinella.levi@polimi.it (M.L.)
2   Department of Design, Politecnico di Milano, Via Durando 38/A, 20158 Milano, Italy
3   Green Coat S.r.l., Strada Romana Nord, 1, 46027 San Benedetto Po, Italy; p.tralli@greencoat.it
*   Correspondence: raffaella.suriano@polimi.it; Tel.: +39-0223993249

Abstract: Fused filament fabrication allows the direct manufacturing of customized and complex products although the layer-by-layer appearance of this process strongly affects the surface quality of the final parts. In recent years, an increasing number of post-processing treatments has been developed for the most used materials. Contrarily to other additive manufacturing technologies, metallization is not a common surface treatment for this process despite the increasing range of high-performing 3D printable materials. The objective of this work is to explore the use of physical vapor deposition sputtering for the chromium metallization of thermoplastic polymers and composites obtained by fused filament fabrication. The thermal and mechanical properties of five materials were firstly evaluated by means of differential scanning calorimetry and tensile tests. Meanwhile, a specific finishing torture test sample was designed and 3D printed to perform the metallization process and evaluate the finishing on different geometrical features. Furthermore, the roughness of the samples was measured before and after the metallization, and a cost analysis was performed to assess the cost-efficiency. To sum up, the metallization of five samples made with different materials was successfully achieved. Although some 3D printing defects worsened after the post-processing treatment, good homogeneity on the finest details was reached. These promising results may encourage further experimentations as well as the development of new applications, i.e., for the automotive and furniture fields.

Keywords: 3D printing; prototyping; surface finishing; physical vapor deposition; mechanical properties; composites; fused deposition modeling; surface quality

Citation: Romani, A.; Mantelli, A.; Tralli, P.; Turri, S.; Levi, M.; Suriano, R. Metallization of Thermoplastic Polymers and Composites 3D Printed by Fused Filament Fabrication. *Technologies* **2021**, *9*, 49. https://doi.org/10.3390/technologies9030049

Academic Editors: Salvatore Brischetto and Eugene Wong

Received: 25 April 2021
Accepted: 8 July 2021
Published: 15 July 2021

Publisher's Note: MDPI stays neutral with regard to jurisdictional claims in published maps and institutional affiliations.

1. Introduction

Thermoplastic polymers and their composites exhibit some distinctive properties such as lightweight and corrosion resistance, which represent a significant advantage when compared to other classes of materials, e.g., metals [1,2]. However, the properties provided by metallic coatings are required in many industrial sectors to improve, for example, their aesthetic appearance and abrasion resistance [3,4]. Moreover, metallic properties can be essential for advanced technological applications, such as electronic fields [5]. In all these applications, the possibility of metalizing plastics has paved the way for new applications for polymer-based materials, reducing the costs and combining polymer advantages with metal properties, such as abrasion resistance and conductivity [6].

The main metallization techniques that are currently used are electro-deposition [7], electroless plating [8], spray techniques [3,9], and physical vapor deposition (PVD) [10]. Among these technologies, sputtering PVD can be considered a suitable technique for plastic substrates because low temperatures can be reached during the deposition, according to the depositing materials, thus preventing polymer degradation [11]. Other attractive advantages of sputtering PVD processes are (a) the capability of covering corners uniformly; (b) the possibility to easily deposit metal alloys by a proper choice of the target composition [12]; and (c) the elimination of dangerous chemicals used in other conventional metallization techniques, making this method an environmentally friendly surface treatment [13].

Recently, several studies were performed to explore the opportunity to cover 3D-printed structures with metallic coatings for different purposes, such as the achievement of a good level of conductivity for rapid-prototyped polymers [14], the fabrication of 3D-printed helical and spiral antennas [15,16], and the development of electroluminescent devices [17]. For 3D-printed parts, the presence of a metallic coating enables the improvement of material properties, such as aesthetics, while also changing the user perception of 3D-printed polymers [13,18]. According to the metal deposited to produce the coating, an enhancement of mechanical strength can be achieved, as well as a good level of abrasion resistance and environmental resistance [19].

In this work, the metallization of fused filament fabrication (FFF) components through a sputtering PVD route was explored to assess the level of aesthetic appearance and geometric accuracy obtained after a chromium coating sputtering deposition as a function of different 3D printing materials and parameters used. For this purpose, a finishing torture test 3D model was designed and 3D-printed with thermoplastics and polymer composites. Different thermoplastic materials were selected to explore the capability to metalize different 3D-printable materials with a wide range of characteristics. This will pave the way for the 3D printing of other polymer-based materials and subsequent chromium metallization to improve the surface finishes of additive manufactured products and expand the range of new potential applications.

2. Materials and Methods

The materials for 3D printing were purchased in the form of filaments and used as-is. Polylactic acid (PLA) filament was supplied by Prusa Polymers a.s., Prague, Czech Republic, with the commercial name Prusament PLA (purchased from Prusa Research a.s., Prague, Czech Republic). Acrylonitrile styrene acrylate (ASA) filament was supplied by FormFutura B.V., Nijmegen, Netherlands, with the commercial name ApolloX (local distributor: Conrad Electronic Italia S.r.l., Bollate, Italy). Polycarbonate and acrylonitrile butadiene styrene copolymer (PC-ABS) filaments were supplied by Ciceri de Mondel S.r.l., Ozzero, Italy, with the commercial name FILOALFA® PC-ABS (local distributor: Mega 3D, Cassano d'Adda, Italy). Co-polyester filament made with Amphora HT5300 polymer (Eastman Chemical Company, Kingsport, TN, USA) was supplied by colorFabb B.V., Belfeld, Netherlands, with the commercial name colorFabb_HT, hereinafter called HT-PE (local distributor: Mega 3D, Cassano d'Adda, Italy). Composite co-polyester filament made with Amphora AM1800 polymer (Eastman Chemical Company, Kingsport, TN, USA) and 20 wt% of carbon fibers were supplied by colorFabb B.V., Belfeld, Netherlands, with the commercial name colorFabb XT-CF20, hereinafter called XT-PE-CF20 (local distributor: Mega 3D, Cassano d'Adda, Italy).

Tensile tests were performed by means of a Zwick Roell Z010 (ZwickRoell GmbH & Co. KG, Ulm, Germany) with a 10 kN cell load. For the tests, the ASTM standard test method D3039/D3039M-17 (2017) was adopted [20]. Following the requirements of the standard test method, a testing speed of 2 mm/min was selected, and the specimen dimensions were defined accordingly. The specimens had a constant rectangular cross-section, a width of 10 mm, and a thickness of 2.4 mm. A gripping length of 30 mm was employed, without the use of tabs, and the gauge length was 20 mm. From the standard test method, the length of the specimens was defined according to the following requirements: 2 times the gripping length + 2 times the width + gauge length, resulting in a final length of 100 mm. A total of 5 specimens were tested, and the mechanical properties were evaluated from the stress–strain curve obtained from the tests. Finally, the mean values and the standard error were evaluated.

Differential scanning calorimetry analysis was performed to measure the glass transition temperature (T_g) of the 3D printing filaments. The tests were performed with a Mettler-Toledo DSC/823e (Mettler Toledo, Columbus, OH, USA) in a N_2 atmosphere. The heating ramp was set from -50 °C to 300 °C with a 20 °/min of heating rate, except for HT-PE, which was analyzed with a heating ramp from 25 °C to 300 °C at 20 °/min heating rate.

Tensile test specimens and the 3D model of the torture test were designed using Fusion 360 (Autodesk, San Rafael, CA, USA) CAD software. Complex surfaces of the 3D model were previously designed with the "Grasshopper" plugin of Rhinoceros (Robert McNeel & Associates, Seattle, WA, USA) and afterward integrated into the 3D model with Fusion 360.

Both the tensile specimens and the torture test samples were 3D printed with a Prusa i3 MK3S FDM 3D printer (Prusa Research a.s., Prague, Czech Republic). Tensile specimens Gcodes were created with Ultimaker Cura 3.6.0 slicing software (Ultimaker B.V., Utrecht, The Netherlands). Specimens were produced without perimeters or top and bottom layers. Moreover, 100% infill was set with a raster angle equal to 0° or 90°, compared to the overall length of the specimen. Torture tests Gcodes were created with Slic3r PE 1.41.3 slicing software (Prusa Research a.s., Prague, Czech Republic). Torture tests were 3D printed with 2 perimeters, 4 top and bottom layers, and 20% infill. Other material-specific 3D printing parameters are reported in Table 1. As indicated in Table 1 by the bed adhesion parameter, the 3D printing was performed with both a paper tape and a brim, which is a layer of the material that extends along the print bed from the edges of the 3D prints for PC-ABS, HT-PE, and XT-PE-CF20 samples. The presence of these two factors was needed for these materials to improve the adhesion of the 3D-printed objects to the printing surface and reduce the deformation in the final samples.

Table 1. The 3D printing parameters of the five materials selected for this work.

Material	Nozzle Diameter (mm)	Nozzle T (°C)	Bed T (°C)	Print Velocity (mm/s)	Bed Adhesion
PLA	0.4	215	60	60	none
ASA	0.4	240	90	40	none
PC-ABS	0.4	240	100	50	brim + paper tape
HT-PE	0.4	250	110	35	brim + paper tape
XT-PE-CF20	0.6	255	80	50	brim + paper tape

After visual inspection of the 3D-printed torture test samples, the chromium layer deposition was carried out through a PVD sputtering proprietary process from Green Coat S.r.l., San Benedetto Po, Italy. During this process, a range of temperatures from 20 to 60 °C can be reached. A UV-curable acrylic-based primer (UNILAC UV BC 05 from Cromogenia-Units S.A., Barcelona, Spain) was deposited onto the surface with a thickness of 70–80 μm before the PVD sputtering. Such values of primer thickness were chosen to assure a good leveling of the surface roughness present on the 3D parts. Before the primer application, the samples were cleaned with an isopropyl alcohol moistened cloth followed by a drying time of 10 min at room temperature. The samples were fixed on a rotational jig to assure the coating of all the surfaces of the samples. Firstly, the liquid primer deposition was carried out with anthropomorphic robots, which sprayed the UV-curable solventless primer to the samples with a rotary cup spinning at 25,000 rpm to achieve the atomization of primer. After the application, the primer UV-curing was performed, using 26 mercury vapor lamps with a 10″ bulb and a peak irradiance of 500 mW/cm^2. The total duration of the UV curing was about 2 min. The jig then entered the in-line metallization system where oxygen plasma provided the surface activation before the chromium layer deposition. The plasma treatment was performed for 90 s with an O_2 flow of 700 sccm, an electrode power of 4 kW, and a rotation speed of 5 rpm. A thin layer of chromium was then deposited through the PVD magnetron sputtering process with a power of 10 kW for each cathode. The sputtering process was carried out at a pressure of 1.5×10^{-3} mbar with a constant argon flow of

600 sccm. The total duration of the deposition was 16 min. The expected chromium layer thickness with these parameters is 200 nm. The samples were then supplied by Green Coat S.r.l. for further investigations after the metallization process. The planar specimen surface roughness was measured by means of a laser profilometer (UBM). The roughness tests were performed along five different lines for each planar substrate. The point density of the measure was 500 points/mm, and the roughness values were reported in terms of the root mean square roughness (Rq). The cost analysis was performed and provided by Green Coat S.r.l.

3. Results and Discussion

3.1. Material Characterization

The 3D-printed thermoplastic polymers and composite were characterized to identify possible applications based on the thermal and mechanical properties. DSC analysis was useful to evaluate the maximum working temperature, based on the T_g. To prevent deformations of the 3D objects, thermal transitions during the metallization process are to be avoided. Consequently, higher T_g polymers are less prone to further deformations or warping after the 3D-printing process. In fact, residual thermal stresses are commonly produced during the fused filament fabrication (FFF) process, especially with low-cost desktop 3D printers, and residual thermal stresses may cause deformations or warping. As shown in Table 2, all the materials, except for PLA, have a T_g higher than 60 °C, which approximately is the maximum temperature reached during metallization through PVD sputtering of chromium. Two glass transition temperatures were identified for ASA and PC-ABS, due to the different polymer phases, which are intrinsically present in these materials. Indeed, the $\mathbf{T_{g\_1}}$ and $\mathbf{T_{g\_2}}$ present in the PC-ABS sample can be attributed to the presence of separate phases composed of polycarbonate and poly-acrylonitrile/styrene, respectively. Concerning ASA glass transitions, the $\mathbf{T_{g\_2}}$ is due to the presence of polyacrylonitrile and polystyrene phases. The lower value of T_g, i.e., 83 °C, can be related to the presence of additives aimed at improving the flowing behavior, thermal stability, and interlayer adhesion as reported by the producer [21,22]. Among the investigated materials, PC-ABS and HT-PE showed the highest values of T_g, higher than 100 °C. Consequently, no deformations are expected due to stress relaxation during the metallization process for these materials. On the contrary, 3D parts produced with PLA are prone to deformations and warping since the material T_g can be reached during the metallization process. Regardless, the exact temperature trend during the metallization process is unknown; considerations about the effects of the metallization process on the samples will be later discussed in Section 3.3.

Table 2. Glass transition temperature values, T_g evaluated with DSC analysis for the materials under investigation.

Material	$\mathbf{T_{g\_1}}$ (°C)	$\mathbf{T_{g\_2}}$ (°C)
PLA	55–60 [23]	n/a
ASA	83	112
PC-ABS	113	147
HT-PE	110	n/a
XT-PE-CF20	80	n/a

Table 3 shows the results of the mechanical tests. As described in the Materials and Methods section, specimens for tensile tests were produced with two different raster angles. The parts produced with the FFF process have anisotropic mechanical properties [24,25]. Consequently, the results of the tensile tests of the two sets of specimens are useful for the evaluation of 3D-printed material anisotropy. As shown in Table 3, specimens with a raster angle equal to 0° show overall higher mechanical properties. Excluding PLA specimens, the materials under investigation exhibited a yield point when produced with this raster angle. On the contrary, all the materials showed a brittle behavior with no yield points when a raster angle of 90° was used. Moreover, the strain at break was lower than the strain at yield of the same material 3D printed with a 0° raster angle. This result could be related to different alignments and concentrations of defects produced during the FFF process. Small voids are commonly found between deposited strands of materials [26,27]. Considering specimens with 0° raster angles, the voids will be present between material strands aligned to the tensile direction. Consequently, the strands will sustain the tensile load. Otherwise, with a 90° raster angle, voids between strands will concentrate in transversal planes with respect to the tensile direction, creating weak surfaces between the strands.

Table 3. Materials tensile mechanical properties.

Material	Raster Angle (°)	Elastic Modulus (GPa)	Stress at Yield (MPa)	Strain at Yield (%)	Stress at Break (MPa)	Strain at Break (%)
PLA	0	2.3 [28]	n/a	n/a	48.7 [29]	2.8 [29]
	90	2.0 [28]	n/a	n/a	48.9 [29]	2.7 [29]
ASA	0	2.1 ± 0.3	43.3 ± 0.6	3.1 ± 0.1	28.2 ± 0.9	17.9 ± 4.5
	90	1.9 ± 0.1	n/a	n/a	25.8 ± 1.2	1.5 ± 0.1
PC-ABS	0	2.2 ± 0.2	39.5 ± 0.5	3 ± 0.1	36.2 ± 1.2	174.2 ± 36.1
	90	1.7 ± 0.1	n/a	n/a	14.1 ± 1.2	1.0 ± 0.1
HT-PE	0	1.7 ± 0.2	41.6 ± 0.8	5.4 ± 0.2	31.7 ± 0.5	128.9 ± 24.8
	90	1.6 ± 0.1	n/a	n/a	35.4 ± 2.9	6.9 ± 3.5
XT-PE-CF20	0	9.1 ± 1.9	60.7 ± 0.6	2.4 ± 0.2	55.3 ± 1.3	4.7 ± 0.2
	90	2.9 ± 0.1	n/a	n/a	40.2 ± 0.4	2.6 ± 0.2

PC-ABS and HT-PE, which were 3D printed with a 0° raster angle, showed an outstanding strain at break of higher than 100%, reaching a value of 174% for PC-ABS. Moreover, among the materials investigated in this work, HT-PE showed the highest strain at yield, equal to 5.4%, and the highest strain at break for the 90° raster angle, equal to 6.9%. Regarding the stress at yield and stress at break, XT-PE-CF20 exhibited the highest strength, due to the contribution of carbon fibers, while a decent strain at yield and strain at break was maintained.

3.2. Design of the Finishing Torture Test Samples

A specific 3D model was designed to evaluate the quality of the PVD sputtering metallization on the 3D-printed pieces. The .stl file is publicly available in an open access repository on Zenodo [30].

In particular, this 3D model took inspiration from the "torture tests" for FFF processes, which are specifically designed to calibrate and test the main critical features for 3D printing (i.e., overhangs, bridging). The qualitative comparison of a set of 3D-printed torture test samples allows finding the most suitable slicing settings for good 3D prints with a specific material. Similarly, the 3D model developed in this work was mainly useful for making a qualitative evaluation of the surface finishing after the PVD sputtering metallization. For this reason, this 3D model can be defined as a "finishing torture test sample" since it aims at testing the quality and homogeneity of the metallization on the 3D-printed parts rather than calibrating the slicing parameters of the 3D printing. Nevertheless, constraints both from FFF and PVD processes were considered to design the finishing torture test, such as the nozzle diameter of the 3D printer, or the clamping system of the PVD sputtering industrial apparatus.

The overall preview of the finishing torture test sample is shown in Figure 1a,b. The maximum dimensions of the 3D model ($53 \times 53 \times 53$ mm^3) were defined to limit the 3D printing times of a single piece. The shape is comparable to a cubic-like volume, and several technical features are placed onto the different surfaces. Their dimensions and positions vary in order to verify both the homogeneity of the post-processing treatment and the shape accuracy after the metallization at different conditions. The specific features are visible in Figure 1 and listed here below:

1. Linear engraving with a maximum depth of 3 mm and a minimum dimension of 0.5×5 mm^2 (Figure 1a,e);
2. Letter engraving and embossing with a depth from 0.2 to 0.6 mm and a height of 0.8 mm (Figure 1a,e);
3. Geometrical embossing with a height from 0.2 to 0.6 mm (Figure 1a,b);
4. Staircase effect and overhangs with an angle from 30° to 75° (Figure 1b,d);
5. Spacing features with a distance from 0.5 to 3 mm (Figure 1b,d);
6. Surface embossing and engraving with a maximum surface variation of 4.5 mm and a nominal depth or height of 0.5 mm (Figure 1b,c);
7. Internal cavity with a maximum overhang of 40° (Figure 1a,e);
8. PVD sputtering fixing holes dimensioned according to the PVD sputtering equipment of the industrial site (Figure 1d).

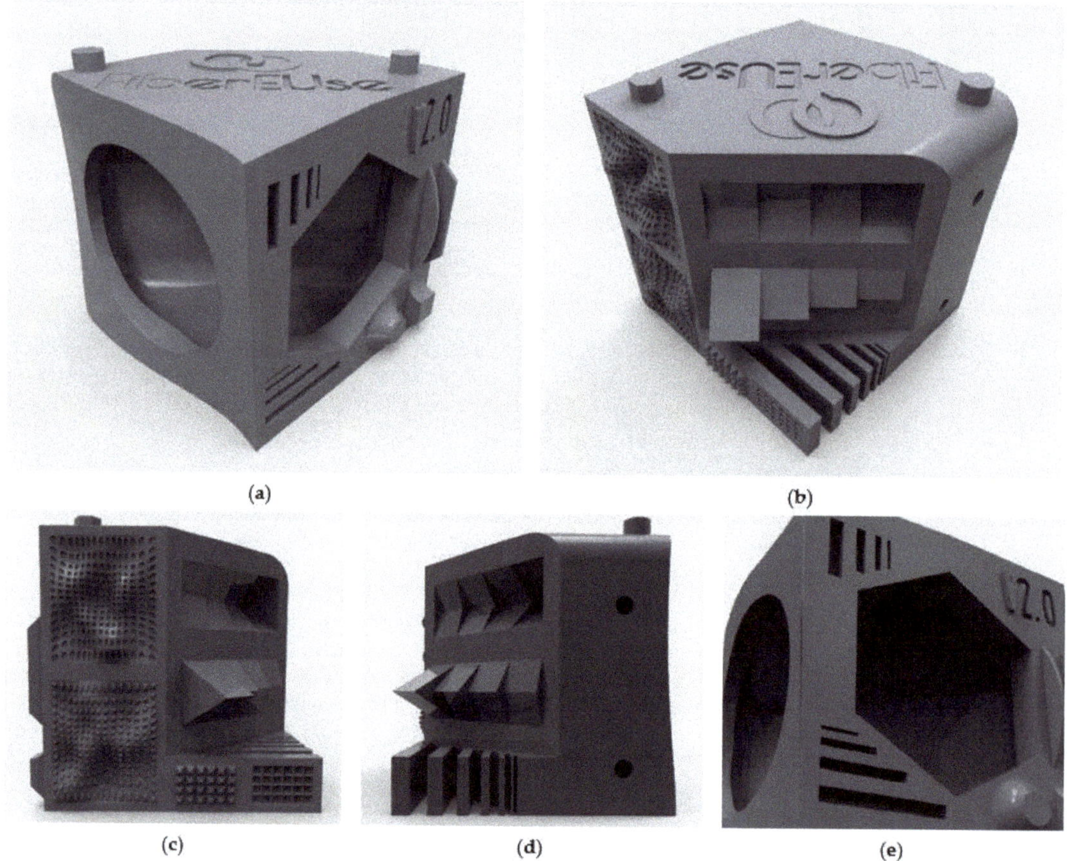

Figure 1. A 3D model of the finishing torture test samples designed in this work (dimensions: $53 \times 53 \times 53$ mm^3): an overview of (**a**) the letter and linear engraving and geometrical embossing surface details; (**b**) the staircase effect and spacing features; (**c**) the surface embossings and engravings; (**d**) the staircase effect, spacing features and the PVD sputtering fixing holes; and (**e**) the internal cavity.

Considering the main purpose, this finishing torture test sample could be suitable for evaluating the surface finishing quality of different post-treatment processes for FFF 3D-printed parts, such as chemical or physical surface treatments [31,32] and other thermal spray technologies. By changing some features, it may be also used to test the surface finishing of 3D-printed surfaces obtained from nonplanar slicing for FFF processes. This novel approach allows the reduction of the staircase effect and the layer-by-layer appearance in nonplanar toolpaths during the fabrication of a 3D-printed part [33–35]. Finally, a similar finishing torture test sample can be designed for other material extrusion 3D-printing technologies as well as new emerging materials.

3.3. The 3D Printing and Metallization of the Finishing Torture Test Samples

Five finishing torture test samples were successfully 3D printed with the parameters and materials previously shown in Table 1, i.e., PLA, ASA, PC-ABS, HT-PE, and XT-PE-CF20. In general, delamination of the layers occurred only for the tests made with ASA and PC-ABS. The most affected areas correspond to the vertical sharp edges and near the internal cavity. Moreover, ASA had also some adhesion problems during the 3D printing, and a raft base was added to the Gcode to improve adhesion to the building plate. No significant superficial defects were detected on the surfaces of the 3D-printed tests, except on the overhang surfaces of the finishing torture test samples made with XT-PE-CF20. At the same time, the surface finishing of this composite material slightly hindered the layer-by-layer appearance of the piece in the z-axis direction, and a random texture can be noticed in the same direction. These aspects are generally noticed in the case of reinforced filaments for two main reasons. Since the presence of fillers highly reduces the printability of these kinds of filaments, a nozzle with a larger diameter is required to avoid clogging inside the hotend, reducing the overall definition of the piece. Moreover, the presence of the filler contributes to modify the surface texture. As a matter of fact, the composite material slightly hinders the layer-by-layer appearance of the piece, thanks to the texture created by the reinforcement particles. Contrastingly, thermoplastic filaments highlight the layer-by-layer appearance. Especially for the ASA filament, the layers are clearly noticeable on the whole finishing torture test sample.

After the PVD sputtering metallization, some considerations can be made by comparing the overall quality of the chromium layer. The quality of the coating is generally higher on planar surfaces without embossed or engraved details. Although the five specimens were successfully coated, several differences can be noticed, according to the different material of the substrate as explained in detail in the following:

1. The PLA sample does not show detachments of the chromium layer, and there are no surface defects due to the metallization. Despite the low T_g, any geometrical deformation is not visible, and good homogeneity was achieved.
2. The delamination of the ASA sample worsened after the metallization. New delaminated points have appeared, and the old ones are enlarged in their dimensions. Moreover, some bubbles can be noticed on the lower surface of the test. These may be due to some entrapped air between the layers, due to the 3D printing and/or PVD sputtering settings. Nevertheless, no detachments and geometrical deformations were detected.
3. For the PC-ABS sample, the delamination has worsened after the PVD sputtering process. The behavior is similar to the ASA sample.
4. The HT-PE sample does not show detachments or superficial defects linked to the metallization or geometrical deformations. A good homogeneity was achieved also in this case.
5. The XT-PE-CF20 sample shows a less shiny surface and lower homogeneity of the chromium layer when compared to the other samples, probably due to the filler. However, no detachment or geometrical deformation was found after the metallization.

In particular, two surface details (surface embossing and letter engraving) and two 3D-printing features (internal cavity and staircase effect) were evaluated through a comparative analysis of the samples. A picture before and after the PVD sputtering metallization was taken for each finishing torture test sample to allow two levels of comparison: (i) the overall quality before and after the coating process; and (ii) the metallization onto the same geometry made with different materials. The comparison of the surface embossing and letter engraving is resumed in the visual matrix of Figure 2. In general, the embossing details were quite visible after the metallization, except for the XT-PE-CF20 test (Figure 2, detail A5). Among the samples under consideration, the highest homogeneity was observed for the PLA sample (Figure 2, detail A1). Contrastingly, the definition of the engraved details worsened in most of the cases. Some inhomogeneity of the chromium layer can be detected in the ASA and PC-ABS samples (Figure 2, details B2 and B3). Furthermore, the layer-by-layer appearance of the PC-ABS and HT-PE tests was highlighted by the metallization (Figure 2, details B3 and B4). In principle, this may be related to the inhomogeneity of the primer layer close to the finest surface details. Further remarks can be made, according to the visual matrix shown in Figure 3. The internal cavity was successfully coated for all the five samples, although the chromium layer of the XT-PE-CF20 test was not homogeneous (Figure 3, Detail A5). Detachments near the cavity are visible in the PC-ABS test (Figure 3, Detail A3). This is also visible beside the detail to evaluate the staircase effect (Figure 3, Detail B3). Some superficial defects were present in this specific area after the metallization of the XT-PE-CF20 sample (Figure 3, Detail B5). From the figures, the staircase effect seems to be highlighted by the metallization, except for the ASA sample (Figure 3, Detail B2).

However, geometrical deformations of the engraved and embossed details were not visible. This result is in agreement with the T_g of the materials previously shown and the temperature range of the PVD sputtering process (20–60 °C). Even for the PLA samples, residual thermal stresses did not significantly affect the final results.

To quantify the quality of the metallized surfaces, the roughness of the sample surfaces was measured before and after the PVD chromium sputtering process. In detail, vertical surfaces were selected, as they better show the typical roughness of the 3D-printing process, hence metallization can show more improvements in those specific areas. As visible in Table 4, the roughness values after the metallization are similar in most cases. The best results were obtained by the HT-PE sample, which also exhibited higher reduction in roughness in terms of root mean square. The XT-PE-CF20 sample showed the highest roughness both before and after the metallization process. In particular, this can be due both to the poorer initial quality related to the use of a larger nozzle (0.6 mm) to avoid fiber-clogging issues, and the presence of carbon fibers, which could be exposed on the surfaces, creating a non-homogeneous substrate for primer and, consequently, for PVD sputtering.

Table 4. Vertical surface roughness values before and after metallization of the finishing torture test samples and their difference.

Material	Rq before PVD(μm)	Rq after PVD (μm)	Rq Reduction (μm)
PLA	7.9 ± 0.5	1.8 ± 3.5	6.1
ASA	11.6 ± 0.7	0.7 ± 0.1	10.9
PC-ABS	6.6 ± 0.6	0.7 ± 0.3	5.9
HT-PE	12.0 ± 1.8	0.3 ± 0.1	11.8
XT-PE-CF20	15.7 ± 2.3	12.0 ± 0.9	3.7

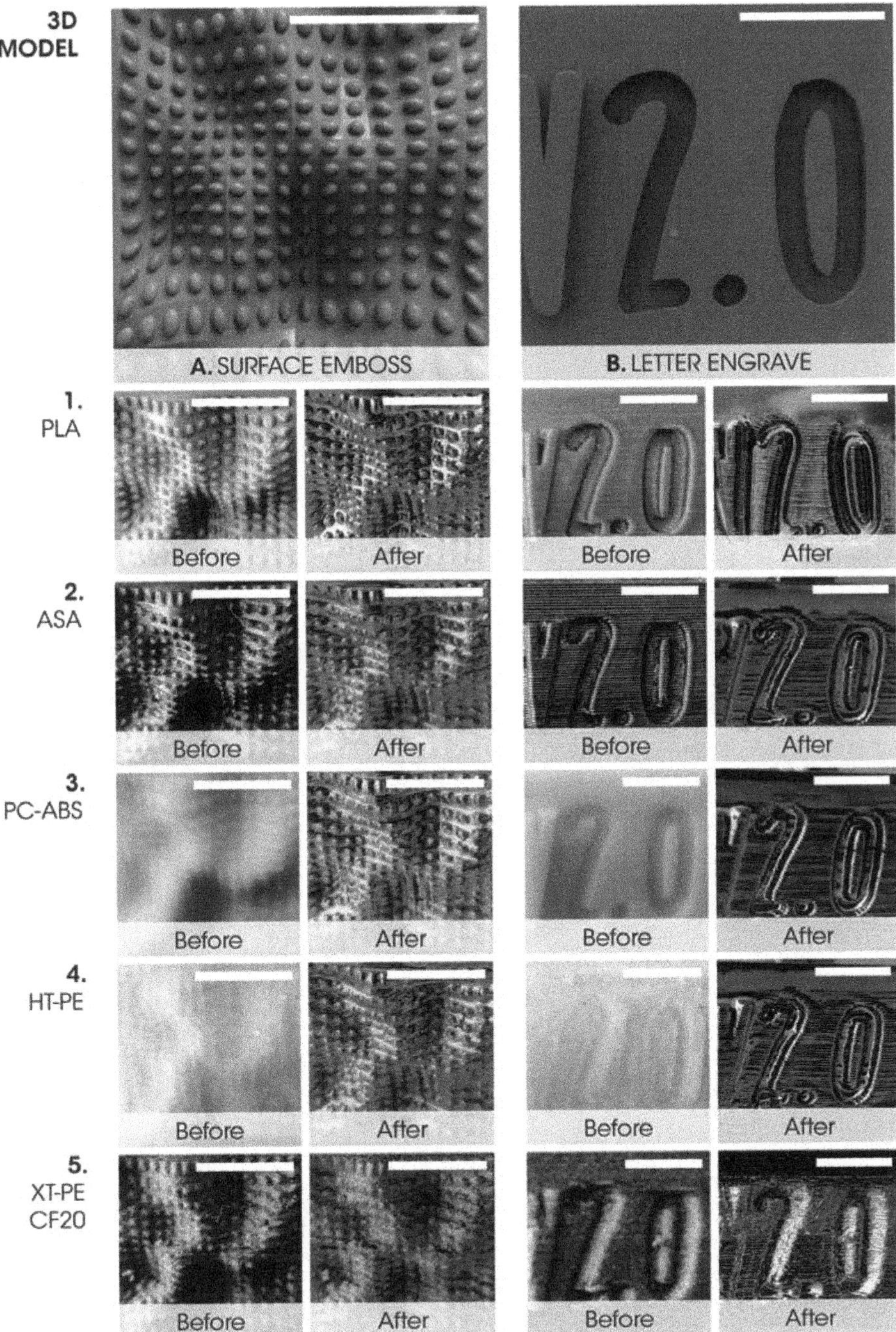

Figure 2. Visual comparative matrix of the (**A**) surface embossed and (**B**) letter engraved details before and after the PVD sputtering metallization of the finishing torture test samples made with (**1**) PLA, (**2**) ASA, (**3**) PC-ABS, (**4**) HT-PE, (**5**) XT-PE-CF20. The white unit bar is equal to 10 mm.

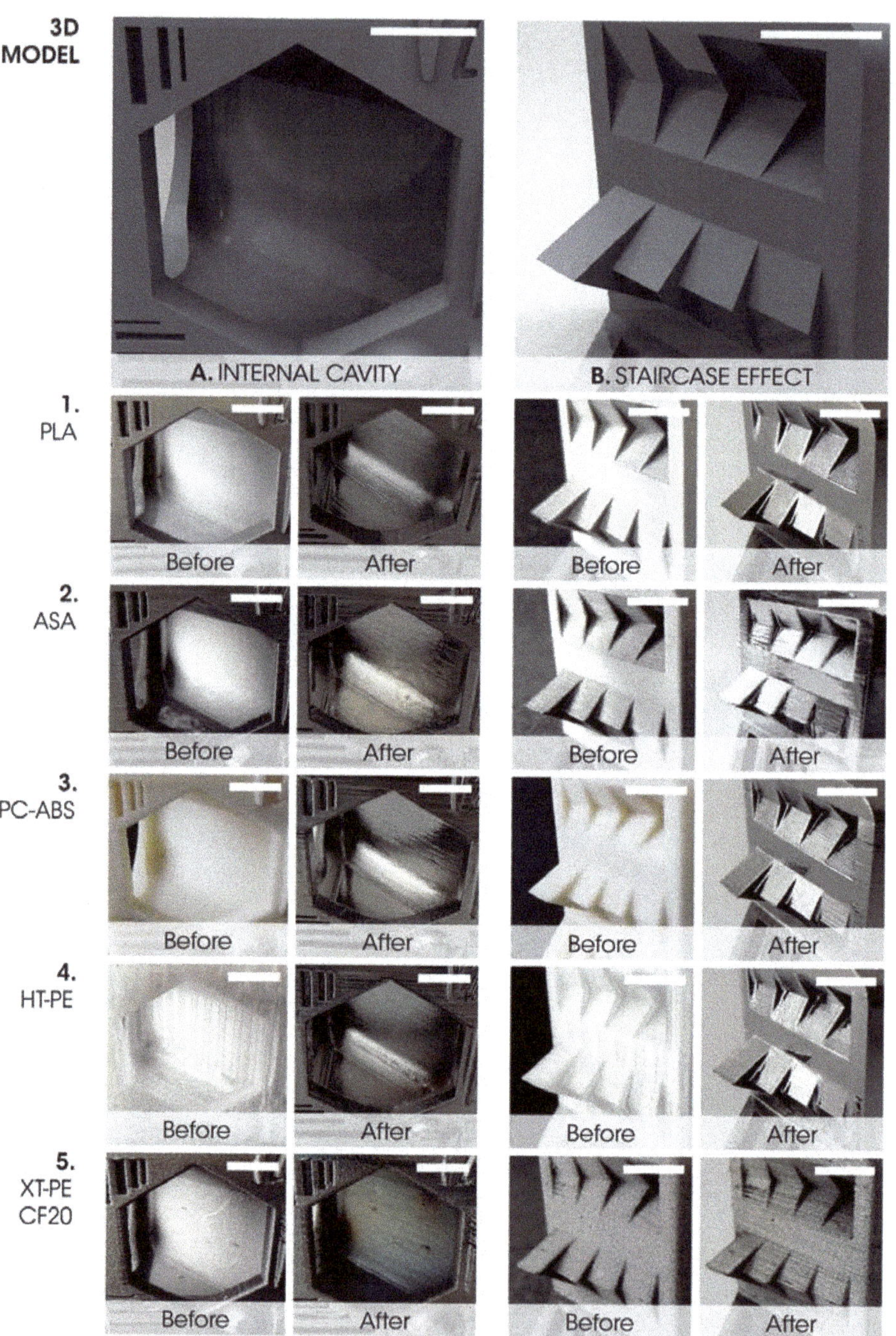

Figure 3. Visual comparative matrix of the (**A**) internal cavity and (**B**) staircase effect before and after the PVD sputtering metallization of the finishing torture test samples made with (**1**) PLA, (**2**) ASA, (**3**) PC-ABS, (**4**) HT-PE, (**5**) XT-PE-CF20. The white unit bar is equal to 10 mm.

3.4. Fields of Application

New applications can be found for 3D-printed thermoplastic and composite materials coated with a metallic layer, although further quantitative tests should be performed to better analyze the PVD sputtering metallization of the selected substrates. As previously mentioned, these kinds of surface coatings were already used for medical and electronic applications with other additive manufacturing processes, i.e., stereolithography [36]. Furthermore, PVD sputtering represents a feasible way to deposit metallic films on FFF 3D-printed substrates, achieving new surface properties [13]. As a result, other technical applications can be exploited after the optimization of the PVD sputtering process, such as in the automotive and furniture fields. New customized and high-performing parts can be fabricated for the automotive industries by taking advantage of both the 3D printing of polymer-based materials and metallization. As a matter of fact, FFF offers a larger range of 3D printable thermoplastics and composite materials when compared to stereolithography as well as the possibility of printing bigger volumes. PVD sputtering could be also used to enhance and change the senso-aesthetic qualities of a specific 3D-printed surface since surface finishing strongly influences not only the performance of a specific product, but also its perception and the emotional response of customers and final users [37]. For instance, designers can highlight or hinder the layer-by-layer appearance of the final product to stimulate a specific response through a proper selection of the substrate material and of the surface finishing. For this reason, customized 3D-printed furniture can benefit from several advantages of metallization not only from a functional perspective, but also from an aesthetic and emotional point of view, i.e., extending the life cycle, protecting the external surfaces, improving the abrasion resistance, and changing the user perception of a 3D-printed product through its finishing. However, aging effects could occur on the chromium/polymer-interfaces, for example, due to redox reactions, occurring at metal/polymer interfaces and chemical reactions caused by the interactions of chromium with carbonyl groups and aromatic rings [38]. For these reasons, future studies will also be dedicated to the investigation of these phenomena. Moreover, PVD sputtering process can be considered an environmentally friendly surface treatment since dangerous chemicals are not used for metal deposition [13].

3.5. Cost Analysis

Finally, to estimate the impact and the cost-efficiency of the proposed metallization process for future industrial exploitations, a cost analysis was performed. More specifically, the costs for the surface treatment and PVD sputtering of torture test samples were determined, including the costs for the coating raw materials, electricity, common gases (such as nitrogen and argon), and periodic maintenance (Table 5).

Table 5. Cost analysis for PVD metallization of a single torture test sample.

Operating Expenditure	Unit (€)
Direct costs of equipment	0.083
Cost of energy (electricity + natural gas)	0.028
Cost of raw materials (e.g., primer)	0.145
Direct labor	0.163
Fixed costs of production	0.054
Cost of expected rejection rate	0.074
General and Administrative expense and Sales	0.109
Capital Expenditure	
Properties, plant, and equipment	0.073
Total Cost	0.729

Considering the results obtained by this cost analysis, it is possible to state that the costs related to energy consumption are approximately 4% of the total cost and have the least impact on the final price. Vice versa, the costs related to the raw material and the operator labor have a greater impact on the total cost (i.e., 20% and 22% of the total cost, respectively). The coating raw materials are usually the highest cost consumables. However, the final cost of the entire metallization process per sample is lower than EUR 1. This means that costs can be measured in cents, for example, for decorative PVD applications on consumer products. Therefore, this PVD sputtering metallization will not add substantial costs to a manufactured part; adding PVD finishing to manufacturing operation can be cost-effective. Moreover, the deposition of a chromium layer enhances the perceived quality of products, because a high-quality product finishing instantly communicates its increased value and influences how customers view manufactured products [39]. For instance, automotive components can be customized by an FFF printing process and then exhibit different senso-aesthetic qualities, thanks to the PVD sputtering metallization. This suggests that, even though the initial investment of a PVD sputtering equipment may be high, it allows the 3D-printed objects to combine customization with good, appealing properties. With that said, in this specific case, the capital expenditure, which represents the cost of buying, maintaining and improving the equipment, is only the 10% of the final cost for the metallic coating of a single object. Furthermore, durable hard coatings can enhance the lifetime of home and office furniture, medical devices, industrial tooling, sporting goods, and many other products, which require being custom-made by 3D printing for fashionable and particular reasons.

4. Conclusions

In this work, the deposition of a chromium layer onto 3D-printed complex shapes was successfully achieved. Five thermoplastic and composite materials were coated through a PVD sputtering process. In detail, the glass transition temperatures and mechanical properties of the selected materials were evaluated not only to preliminary define their PVD processability, but also to showcase the metallization of 3D printable materials with different characteristics. Considering the maximum temperatures reached during PVD, only PLA parts should have been affected by deformations linked to residual thermal stresses. Nevertheless, no visible deformations were found on the PLA torture test sample after the metallization.

A finishing torture test sample was specifically designed to test metallization onto different features that are commonly fabricated through FFF processes. Additionally, this 3D model may be useful to evaluate (i) the quality of different surface finishing for 3D printing; (ii) different 3D-printable materials for FFF; and (iii) other material extrusion 3D-printing processes and slicing approaches after some changes to the main features. Afterward, the 3D printing and metallization of the torture tests were successfully achieved with all the selected materials. In general, good homogeneity was obtained with PLA and HT-PE, whereas the metallization onto the other materials was mainly affected by the presence of 3D-printing defects. Among the materials under investigation, the HT-PE sample exhibited the lowest values of roughness after the PVD sputtering process and the highest reduction of roughness induced by the metallization. Moreover, neither detachments nor geometrical deformations and surface defects due to the metallization process were observed for HT-PE. Especially for applications requiring high thermal stability and good mechanical properties, HT-PE samples can be an appropriate choice, due to the high values of the glass transition temperature and good quality of the PVD sputtered surfaces. However, any detachment of the chromium layer or geometrical deformation was not observed for the carbon-based composite (XT-PE-CF20), and promising results were also achieved in this case.

Considering the different properties of the materials shown in this work, a wider range of new applications can be developed, i.e., in the automotive and furniture sectors. As a matter of fact, this finishing treatment can influence not only the technical properties of new products, but also the user perception. Further quantitative tests should be performed to foster the use of PVD for 3D-printed thermoplastics and composites, such as assessing the abrasion and adhesion resistance of the coatings. Nevertheless, these promising results will potentially allow the exploitation of 3D-printed thermoplastics and composites for the design of new products.

Author Contributions: Conceptualization, A.R., A.M. and R.S.; software, A.R.; validation, A.R. and A.M.; formal analysis, A.R., A.M. and P.T.; investigation, A.R., A.M. and P.T.; resources, A.R. and A.M.; data curation, A.R. and A.M.; writing—original draft preparation, A.R., A.M. and R.S.; writing—review and editing, A.R., A.M., P.T., S.T., M.L. and R.S.; visualization, A.R.; supervision, S.T., M.L. and R.S.; project administration, S.T., M.L. and R.S.; funding acquisition, S.T. and M.L. All authors have read and agreed to the published version of the manuscript.

Funding: This research has received funding from the European Union's Horizon 2020 research and innovation program, under the grant agreement No. H2020-730323-1. The present work is part of the FiberEUse project, entitled "Large Scale Demonstration of New Circular Economy Value-chains based on the Reuse of End-of-life reinforced Composites".

Institutional Review Board Statement: Not applicable.

Informed Consent Statement: Not applicable.

Data Availability Statement: Publicly available data sets were analyzed in this study. The data can be found here: [https://github.com/piuLAB-official/Dataset_A.Romani_2021_Technologies (accessed on 25 April 2021)]. If these data are used, please cite them in the following way: [data set] Alessia Romani, Andrea Mantelli, Stefano Turri, Marinella Levi and Raffaella Suriano. 2021. Metallization of thermoplastic polymers and composites 3D-printed by Fused Filament Fabrication; https://github.com/piuLAB-official/Dataset_A.Romani_2021_Technologies (accessed on 25 April 2021).

Acknowledgments: The authors would like to thank GreenCoat S.r.l. for the PVD sputtering metallization performed on the torture test samples.

Conflicts of Interest: The authors declare no conflict of interest.

References

1. Manoj Prabhakar, M.; Rajini, N.; Ayrilmis, N.; Mayandi, K.; Siengchin, S.; Senthilkumar, K.; Karthikeyan, S.; Ismail, S.O. An overview of burst, buckling, durability and corrosion analysis of lightweight FRP composite pipes and their applicability. *Compos. Struct.* **2019**, *230*, 111419. [CrossRef]
2. Liu, M.; Guo, Y.; Wang, J.; Yergin, M. Corrosion avoidance in lightweight materials for automotive applications. *Mater. Degrad.* **2018**, *2*, 24. [CrossRef]
3. Moridi, A.; Hassani-Gangaraj, S.M.; Guagliano, M.; Dao, M. Cold spray coating: Review of material systems and future perspectives. *Surf. Eng.* **2014**, *30*, 369–395. [CrossRef]
4. Electroless Nickel Coating. Available online: http://www.starmetal-tr.com/en/electroless_nickel.html (accessed on 17 April 2021).
5. Ghosh, S. Electroless copper deposition_ A critical review. *Thin Solid Films* **2019**, *669*, 641–658. [CrossRef]
6. Oliveira, F. A new approach for preparation of metal-containing polyamide/carbon textile laminate composites with tunable electrical conductivity. *J. Mater. Sci.* **2018**, *53*, 11444–11459. [CrossRef]
7. Lammel, P.; Whitehead, A.H.; Simunkova, H.; Rohr, O.; Gollas, B. Droplet erosion performance of composite materials electroplated with a hard metal layer. *Wear* **2011**, *271*, 1341–1348. [CrossRef]
8. Schaubroeck, D. Surface modification of an epoxy resin with polyamines and polydopamine: Adhesion toward electroless deposited copper. *Appl. Surf. Sci.* **2015**, *353*, 238–244. [CrossRef]
9. Gonzalez, R.; Ashrafizadeh, H.; Lopera, A.; Mertiny, P.; McDonald, A. A Review of Thermal Spray Metallization of Polymer-Based Structures. *J. Therm. Spray Technol.* **2016**, *25*, 897–919. [CrossRef]
10. Maurer, C.; Schulz, U. Solid particle erosion of thick PVD coatings on CFRP. *Wear* **2014**, *317*, 246–253. [CrossRef]
11. Baptista, A.; Silva, F.J.G.; Porteiro, J.; Míguez, J.L.; Pinto, G.; Fernandes, L. On the Physical Vapour Deposition (PVD): Evolution of Magnetron Sputtering Processes for Industrial Applications. *Procedia Manuf.* **2018**, *17*, 746–757. [CrossRef]
12. Constantin, R.; Miremad, B. Performance of hard coatings, made by balanced and unbalanced magnetron sputtering, for decorative applications. *Surf. Coat. Technol.* **1999**, *120–121*, 728–733. [CrossRef]
13. White, J. Environmentally benign metallization of material extrusion technology 3D printed acrylonitrile butadiene styrene parts using physical vapor deposition. *Addit. Manuf.* **2018**, *22*, 279–285. [CrossRef]
14. Dixit, N.K.; Srivastava, R.; Narain, R. Electroless Metallic Coating on Plastic Parts Produced by Rapid Prototyping Technique. *Mater. Today Proc.* **2017**, *4*, 7643–7653. [CrossRef]
15. Ghassemiparvin, B.; Ghalichechian, N. Design, fabrication, and testing of a helical antenna using 3D printing technology. *Microw. Opt. Technol. Lett.* **2020**, *62*, 1577–1580. [CrossRef]

16. Filonov, D.; Kolen, S.; Shmidt, A.; Shacham-Diamand, Y.; Boag, A.; Ginzburg, P. Volumetric 3D-Printed Antennas, Manufactured via Selective Polymer Metallization. *Phys. Status Solidi RRL Rapid Res. Lett.* **2019**, *13*, pssr.201800668. [CrossRef]
17. Brubaker, C.D.; Newcome, K.N.; Jennings, G.K.; Adams, D.E. 3D-Printed alternating current electroluminescent devices. *J. Mater. Chem. C* **2019**, *7*, 5573–5578. [CrossRef]
18. Bernasconi, R.; Natale, G.; Levi, M.; Magagnin, L. Electroless Plating of PLA and PETG for 3D Printed Flexible Substrates. *ECS Trans.* **2015**, *66*, 23–35. [CrossRef]
19. Afshar, A.; Mihut, D. Enhancing durability of 3D printed polymer structures by metallization. *J. Mater. Sci. Technol.* **2020**, *53*, 185–191. [CrossRef]
20. *ASTM D3039/D3039M-17*; Standard Test Method for Tensile Properties of Polymer Matrix Composite Materials; ASTM International: West Conshohocken, PA, USA, 2017.
21. *Polymer Data Handbook*; Mark, J.E. (Ed.) Oxford University Press: New York, NY, USA, 1999; ISBN 978-0-19-510789-0.
22. ApolloX. Available online: https://www.formfutura.com/shop/product/apollox-2779 (accessed on 24 June 2021).
23. TDS Prusament PLA. Available online: https://prusament.com/media/2018/07/PLA_TechSheet_ENG22052020.pdf (accessed on 17 April 2021).
24. Zaldivar, R.J.; Witkin, D.B.; McLouth, T.; Patel, D.N.; Schmitt, K.; Nokes, J.P. Influence of processing and orientation print effects on the mechanical and thermal behavior of 3D-Printed ULTEM® 9085 Material. *Addit. Manuf.* **2017**, *13*, 71–80. [CrossRef]
25. Dey, A.; Yodo, N. A Systematic Survey of FDM Process Parameter Optimization and Their Influence on Part Characteristics. *J. Manuf. Mater. Process.* **2019**, *3*, 64. [CrossRef]
26. Cicala, G.; Giordano, D.; Tosto, C.; Filippone, G.; Recca, A.; Blanco, I. Polylactide (PLA) Filaments a Biobased Solution for Additive Manufacturing: Correlating Rheology and Thermomechanical Properties with Printing Quality. *Materials* **2018**, *11*, 1191. [CrossRef]
27. Lanzotti, A.; Grasso, M.; Staiano, G.; Martorelli, M. The impact of process parameters on mechanical properties of parts fabricated in PLA with an open-source 3-D printer. *Rapid Prototyp. J.* **2015**, *21*, 604–617. [CrossRef]
28. Yao, T.; Ye, J.; Deng, Z.; Zhang, K.; Ma, Y.; Ouyang, H. Tensile failure strength and separation angle of FDM 3D printing PLA material: Experimental and theoretical analyses. *Compos. Part B Eng.* **2020**, *188*, 107894. [CrossRef]
29. Hanon, M.M.; Marczis, R.; Zsidai, L. Influence of the 3D Printing Process Settings on Tensile Strength of PLA and HT-PLA. *Period. Polytech. Mech. Eng.* **2020**, *65*, 38–46. [CrossRef]
30. Romani, A.; Mantelli, A. FFF Finishing Torture Test for PVD chromium metallization. *Zenodo* **2021**. [CrossRef]
31. Kuo, C.-C.; Chen, C.-M.; Chang, S.-X. Polishing mechanism for ABS parts fabricated by additive manufacturing. *Int. J. Adv. Manuf. Technol.* **2017**, *91*, 1473–1479. [CrossRef]
32. Adel, M. Polishing of fused deposition modeling products by hot air jet_ Evaluation of surface roughness. *J. Mater. Process. Tech* **2018**, 73–82. [CrossRef]
33. Ahlers, D.; Wasserfall, F.; Hendrich, N.; Zhang, J. 3D Printing of Nonplanar Layers for Smooth Surface Generation. In Proceedings of the IEEE 15th International Conference on Automation Science and Engineering (CASE), Vancouver, BC, Canada, 22–26 August 2019; pp. 1737–1743.
34. Elkaseer, A.; Müller, T.; Rabsch, D.; Scholz, S.G. Impact of Nonplanar 3D Printing on Surface Roughness and Build Time in Fused Filament Fabrication. In *Proceedings of the Sustainable Design and Manufacturing 2020*; Scholz, S.G., Howlett, R.J., Setchi, R., Eds.; Springer: Singapore, 2021; pp. 285–295.
35. Nisja, G.A.; Cao, A.; Gao, C. Short review of nonplanar fused deposition modeling printing. *Mater. Des. Process. Commun.* **2021**. [CrossRef]
36. Xiao, R.; Feng, X.; Fan, R.; Chen, S.; Song, J.; Gao, L.; Lu, Y. 3D printing of titanium-coated gradient composite lattices for lightweight mandibular prosthesis. *Compos. Part B Eng.* **2020**, *193*, 108057. [CrossRef]
37. Chen, X.; Shao, F.; Barnes, C.; Childs, T.; Henson, B. Exploring relationships between touch perception and surface physical properties. *Int. J. Des.* **2009**, *3*, 67–76.
38. Friedrich, J.F.; Unger, W.E.S.; Lippitz, A.; Koprinarov, I.; Kühn, G.; Weidner, S.; Vogel, L. Chemical reactions at polymer surfaces interacting with a gas plasma or with metal atoms—their relevance to adhesion. *Surf. Coat. Technol.* **1999**, *116–119*, 772–782. [CrossRef]
39. Zuo, H. The Selection of Materials to Match Human Sensory Adaptation and Aesthetic Expectation in Industrial Design. *Metu J. Fac. Archit.* **2010**, *27*, 301–319. [CrossRef]

MDPI

St. Alban-Anlage 66

4052 Basel

Switzerland

Tel. +41 61 683 77 34

Fax +41 61 302 89 18

www.mdpi.com

Technologies Editorial Office

E-mail: technologies@mdpi.com

www.mdpi.com/journal/technologies